Environment a

The increasing awareness of the human impact on the environment is having a profound effect on the concept and content of citizenship – one of the fundamental institutions that structures human relations. In what is the first introduction of its kind, this book provides an accessible, stimulating, and multidimensional overview of the many ways in which concern for the environment – driven primarily by the preoccupation with sustainability – is reshaping our understanding of citizenship.

Environment and Citizenship is structured into three parts. Part I introduces the reader to the concept and theories of citizenship and explores the impact that environmental concerns is having on contemporary formulations of citizenship, both traditional (e.g. national, liberal, and republican) and emerging (e.g. cosmopolitan, ecological, and ecofeminist). Part II explores the practical manifestations of environmental citizenship, with each chapter focusing on a particular actor: citizens, governments and corporations. These chapters include references to examples and case studies from a wide range of countries, broadly categorized as belonging to the Global North and the Global South. Part III explores the making of green citizens and outlines the dominant articulations of environmental citizenship that emerge from formal education, news media, and popular culture. The book concludes with a general reflection on the present and future of environmental citizenship.

The book contains a variety of illustrations, boxed case-studies, links to online resources, and suggestions for further reading. This original and engaging text is essential reading for students and scholars of environmental politics, sustainability studies, and development studies, as well as for environmental activists, policy practitioners, and environmental educators. More broadly, this book will appeal to anyone interested in and concerned with issues of sustainability, social justice, and citizenship in the twenty-first century.

Benito Cao is Lecturer in Politics at the University of Adelaide, Australia. His main area of research focuses on the intersection of politics and popular culture, in particular, the articulations of this intersection in the formation and formulation of colonial, imperial, and national identities. He has received numerous teaching awards and has published in the fields of pedagogy, cultural studies, identity politics, and environmental activism in Brazil.

'It is hard to speak too highly of this marvellous book. *Environment and Citizenship* is at once a readable, entertaining, and fully comprehensive introduction to the topic, and a contribution to original scholarship. Enlivened by examples, and leavened with online resources and student activities, this book will be an indispensable companion to anyone teaching or researching in the field.'

Professor Andrew Dobson, Keele University, UK, and author of *Citizenship and the Environment* (2003)

'Cao's effort to chart the emergence of the environmental citizen reaches deep into the historical and conceptual roots of both citizenship and the politics of nature. His sustained effort to grapple with theoretical debates in the field is matched by a rich engagement with the realm of practice, shedding light on the ways that individuals, social movements, governments, and corporations contribute to competing conceptions of environmental citizenship. Spanning diverse perspectives across both social and geographical difference, the text provides a thought-provoking read for students and instructors alike.'

Professor Alex Latta, Wilfrid Laurier University, Canada, and co-editor (with Hannah Wittman) of *Environment and Citizenship in Latin America: Natures, Subjects and Struggles* (2012)

'Cao's *Environment and Citizenship* makes an important contribution in terms of the maturing of the complex debate between green political theory and its concerns and the dynamics and evolution of citizenship. The tripartite focus of the book on theories (Part I), actions (Part II), and representations (Part III) of the relationship/s between environment and citizenship produces some novel, interesting, and provocative ideas. *Environment and Citizenship* shows how green values and especially global ecological problems such as climate change challenge citizenship, such as its nation-state focus or indeed the automatic and exclusive association of citizenship with human beings. The book is clearly written and makes good use of practical examples and case studies to illustrate normative arguments. Thus it is a perfect textbook, one that deserves to be widely used.'

Professor John Barry, Queens University Belfast, Northern Ireland, and author of *The Politics of Actually Existing Unsustainability* (2012)

Routledge Introductions to Environment Series
Published and Forthcoming Titles

Environmental Science texts

Atmospheric Processes and Systems
Natural Environmental Change
Environmental Biology
Using Statistics to Understand The
Environment
Environmental Physics
Environmental Chemistry
Biodiversity and Conservation, 2nd
Edition
Ecosystems, 2nd Edition
Coastal Systems 2nd Edition

Series Editor:
Timothy Doyle

Environment and Society texts

Environment and Philosophy
Energy, Society and Environment, 2nd
Edition
Gender and Environment
Environment and Business
Environment and Law
Environment & Society
Representing the Environment
Sustainable Development
Environment and Social Theory, 2nd
Edition
Environmental Values
Environment and Politics, 3rd Edition
Environment and Tourism, 2nd Edition
Environment and the City
Environment, Media and
Communication
Environmental Policy, 2nd Edition
Environment and Economy
Environment and Food
Environmental Governance
Environment and Citizenship

Forthcoming Titles

Sustainable Development, 2nd Edition
Environment and Politics, 4th Edition

Environment and Citizenship

Benito Cao

Routledge
Taylor & Francis Group

LONDON AND NEW YORK

First published 2015
by Routledge
2 Park Square, Milton Park, Abingdon, Oxon OX14 4RN

and by Routledge
711 Third Avenue, New York, NY 10017

Routledge is an imprint of the Taylor & Francis Group, an informa business

© 2015 Benito Cao

British Library Cataloguing in Publication Data
A catalogue record for this book is available from the British Library

Library of Congress Cataloging in Publication Data
A catalog record for this book has been requested

ISBN: 978-0-415-63779-4 (hbk)
ISBN: 978-0-415-63780-0 (pbk)
ISBN: 978-0-203-08433-5 (ebk)

Typeset in Times New Roman
by Keystroke, Station Road, Codsall, Wolverhampton

Printed and bound in the United States of America by Publishers Graphics, LLC on sustainably sourced paper.

Contents

Figures

Boxes

Preface to Routledge Introductions to Environment Series

Series Editor: Timothy Doyle,

Professor of Politics and International Studies, University of Adelaide, Australia;

Professor of Politics and International Relations, Keele University, United Kingdom;

Distinguished Research Fellow, Curtin University, Australia;

Chair, Indian Ocean Rim Association Academic Group, Port Louis, Mauritius;

Chief Editor, *Journal of the Indian Ocean Region,* Routledge, Taylor & Francis, London, UK.

It is openly acknowledged that a critical understanding of socioeconomic, political and cultural processes and structures is central in understanding environmental problems and establishing alternative modes of equitable development. As a consequence, the maturing of environmentalism has been marked by prolific scholarship in the social sciences and humanities, exploring the complexity of society–environment relationships.

This series builds on the work of the founding series editor, David Pepper, to continue to provide an understanding of the central socioeconomic, political and cultural processes relating to environmental studies, providing an interdisciplinary perspective to core environmental issues. David initiated the series by celebrating the close connections between the academic traditions of environmental studies and the emergence of the green movement itself. Central to the goals of the movement were social and environmental change. As the 'new science' of ecology was interdisciplinary, seeking to understand relationships within and between ecosystems, so too was the belief within the academy (informed by the movement) that real environmental change could only emerge if traditional borders and boundaries of knowledge and power were bypassed, transgressed and, where necessary, challenged.

This bid for engaged knowledge and interdisciplinarity also informs the structure and 'pitch' of these books. For it is no good communicating with just one particular group within society. It is equally important to construct forms of knowledge which can cross over demographic and market borders, bringing together communities of people who may never 'meet' in the usual course of events. Thus, the epistemological design of this series is oriented around three particular audiences, providing an unparalleled interdisciplinary perspective on key areas of environmental study: (1) students (at undergraduate and coursework postgraduate levels); (2) policy practitioners (in civil society, governments and corporations); and (3) researchers. It is important to note, therefore, that these books – though strongly used in diverse levels of tertiary teaching – are also built, in large part, on the primary and often ground-breaking research interests of the authors.

In his own ground-breaking work, David Pepper was particularly interested in exploring the relationships between capitalism, socialism and the environment. David argued that the modern environmentalist movement grew at a rapid pace in the last third of the twentieth century. It reflected popular and academic concerns about the local and global degradation of the physical environment which was increasingly being documented by scientists. It soon became clear, however, that reversing such degradation was not merely a technical and managerial matter: merely knowing about environmental problems did not of itself guarantee that governments, businesses or individuals would do anything about them. Since David wrote his last series Preface, this focus has continued to be important, but with special permutations as time has worn on. One more recent, key feature of these society–environment relationships has been the clear differentiation between the environmentalisms of the majority worlds (the Global South) and the environmentalisms of the minority worlds (the more affluent Global North). Wherein environmentalism came to the less affluent world later (in the 1980s), key environmental leadership is now being provided by activists in the South, oriented around a postcolonial environmentalism: with its key issues of human dispossession and survival: water, earth (food security and sustainability), fire (energy), and air (not climate). Much of the focus in environmentalism in the South relates to a critique of capitalism and its big business advocates as being the major perpetrators of severe environmental problems which confront the Earth. In the Global North (where the modern movement began in the 1960s), there has been far more emphasis on postmaterialism and postindustrialism and, more recently, building a *sustainable capitalism.*

Climate change is now the neoliberal *cause célèbre* of this approach, with its heavily biased focus on market mechanisms and green consumerism as answers to environmental crises. In fact, climate change, in the Global North, has now become so powerful and omnipresent that many more affluent-world green activists and academics now comprehend *all* environmental problems within its rubric, its story. Of course, climate change issues will continue to be crucial to the planet's continued existence, but more importantly, it must be acknowledged that in living social movements – like the green movement – issues will come and go; will be re-ordered and re-arranged on the issue attention cycle; be re-badged under different symbols, signs and maps; and new green narratives, issues and stories will emerge. The environment movement, born in the North – and its associated academic studies – will continue to be the foremost global social movement for change for many years to come – if it can continue to truly engage with the Global South – utilising these new and revised banners, issues and colours to continually and creatively mark out its territories, constructing versions of environmentalism for all; not just for the few. And it is within these new sites of politics and knowledge where some of the most exciting advances in the relationships between societies and 'nature' will continue to emerge and be celebrated. Much still is to be learned from our universe, the planet Earth, its human and nonhuman communities.

Tim Doyle
October 2014

Acknowledgements

I dedicate this book to my wife and life partner, Adela. This book would not exist without her research, administrative, editorial and intellectual support. Her critical insights and editorial skills have been crucial throughout the whole project, and the final product owes much to her ability to identify what was clear and what needed clarity, what worked and what needed work, what was interesting and what was unnecessary. The text has also benefitted hugely from the generous and constructive comments by the three anonymous reviewers of the initial manuscript. The final product is better because of the additional work done after their insightful suggestions. Of course, whatever errors or deficiencies remain in this work are my sole responsibility.

I am deeply grateful to the series editor, and top bloke (as we say in Australia), Professor Timothy Doyle. It was Tim who approached me with the idea of writing a book for this series. His confidence in my ability to pull this off made the decision easy and I hope that confidence has been fully repaid. I am also deeply grateful to the Routledge team, particularly Andrew Mould, Faye Leerink and Sarah Gilkes. The work that Andrew and those who reviewed the book proposal did in the early stage of the project was crucial in shaping the text, and the outcome is all the better for their effort. Faye and Sarah were exemplary in their administrative support and email exchanges over the months it has taken to get to this point.

I also wish to thank the students I have taught over the years. Their comments and questions have challenged me to explain and explore the content I teach with as much clarity as possible, to identify what is essential and what needs unpacking, and to map concepts, theories and the key content of all the courses I teach. That challenge is what drives me to make sense of things in ways that can be clearly communicated, so that students, I and others (and not only a handful of specialists) can understand the important issues that occupy academics, both as teachers and researchers.

Finally, I would like to acknowledge and thank all the individuals and organizations who have granted me permission to use their work (photos, images, cartoons). Not only does the book look better for that but those images help to illustrate important issues, concepts and themes. I wish to give a special thanks to everyone who places their content on Creative Commons. Their generosity makes the world a much better place. Individual thanks go to: Cathy Wilcox (Figure 1.3 and Figure 6.2), Seán Mullarkey (Figure 4.5 and Figure 7.1), Stacey Griffiths and Carmel College (Figure 3.1), Oxford University Press (Figure 3.3), Martin Rowe (Figure 4.4), Andrew Weldon (Figure 5.4), Lizzie Wilberforce (Figure 6.3), Chris Madden (Figure 6.4), and Colin Beavan (Figure 7.5).

And, last but not least, to the 'animal citizen' that has brightened my (and Adela's) domestic life during the two years it has taken to complete this project, our gorgeous ginger tabby, Bongo . . . who in recent months we have begun referring to as 'Citizen Bongo'.

Introduction

> The map is not the territory.
>
> Alfred Korzybski (1931)

To the extent that human life is dependent on it, the environment has always been central to humanity. However, it is only in the past decades, and particularly since the turn of the century, that we find references to the state of the environment everywhere we turn, often telling us we need to do something about it (and fast!). This growing preoccupation with the environment has overlapped with the renewed interest in citizenship that began in the 1990s. Since then, citizenship has become a hot topic and a concept whose complexity and flexibility have increased dramatically, producing a significant number of new theoretical articulations (e.g. environmental citizenship) that challenge and coexist with traditional ones (e.g. republican citizenship).

This overlap of environment and citizenship is not just happening at the theoretical level. In the past two decades, the language of citizenship has been used to frame environmental issues, in particular, concerns with sustainability. Indeed, it has become almost obligatory to mention references to rights and duties to address all matters environmental. Green movements and organizations, governments, and private corporations regularly invoke citizenship to make environmental claims and inform environmental actions, policies and initiatives. In addition, education for citizenship and education for the environment have been on a gradual path of convergence, further contributing to frame the environment as a matter of citizenship. Last but not least, the media increasingly are drawing on the language of citizenship to frame their coverage of environmental issues.

This convergence of environment and citizenship has led to the emergence of a new field of study, commonly known as environmental citizenship. As a field of study, environmental citizenship is essentially a subfield of environmental politics, and is closely related to the rethinking of the relationship between ecology and democracy brought about by the influence of green political thought (e.g. Dobson 2007; Eckersley 2004). However, since the so-called 'return of the citizen' in the 1990s, environmental citizenship has become a significant field in its own right. Indeed, there is a growing literature on environmental citizenship – variously referred to as 'environmental', 'ecological', 'green' and 'sustainable' citizenship. This literature includes hundreds of articles and a growing number of monographs (e.g. Dobson 2003; Dobson and Bell 2006; MacGregor 2006a; Smith and Pangsapa 2008).

The aim of this text is to map and explore this field and, more broadly, the various intersections between environment and citizenship. But before we proceed with that task, it is important to keep in mind that the territory of environmental citizenship is far richer, more complex and dynamic than any map can ever convey. We must also keep in mind that maps, as well as guides to navigate a particular territory, are snapshots and representations of that territory, not *the* territory. Territories, especially social ones, are always evolving. Their mapping is always incomplete, an unfinished project that must adjust to the changing terrain. Moreover, mapping is simultaneously a process of discovery (of what is out there) and imagination (of what it means). Its interpretative nature means that the distinctions and categories on which mapping relies are always contested. They are little more than markers, labels or symbols that enable us to make sense of the complex and dynamic configuration of a given territory. In this case, the territory is the relationship between citizenship and the environment, and the field of environmental citizenship that has emerged to explore and make sense of that relationship in its various manifestations.

The starting point of the text is the notion that environmental concerns and ecological values are transforming citizenship in ways that resemble a revolution at work, with the potential to radically transform what it means to be a citizen in the twenty-first century. This argument follows from the work of Andrew Dobson, and is neatly captured in the cover of his seminal book, *Citizenship and the Environment* (2003). The cover is a variation on the iconic painting of the French Revolution, Eugène Delacroix's *Liberty Leading the People* (1830). The significance of the cover lies in the fact that this was the revolution that signalled the definitive birth of modern citizenship, the 'birth certificate' of which is the Declaration of the Rights of Man and the Citizen (1789). The variation of the painting is simple but highly effective: instead of the French national flag, Liberty leads the people holding a tree branch. See Figure 3.3 on p. 86. The cover suggests that citizenship is undergoing a revolutionary transformation, similar to the one it underwent during the French Revolution. This transformation is informed by environmental concerns and ecological values (symbolized by the tree branch). Dobson's book is dedicated to exploring this revolution. His work is given special attention in Chapter 3.

The question of whether citizenship is undergoing a revolution is an important thread throughout this text. This translates into a series of related questions: are we witnessing the emergence of something radically new (as the term revolution suggests)? Will environmental concerns with sustainability and the integration of ecological values bring about the end of citizenship as we know it? Or can these values and concerns be accommodated within mainstream notions of citizenship? Is environmental citizenship simply another variation of a concept that has remained essentially unchanged (in its fundamental architecture) for centuries? Can environmental conceptions of citizenship be accommodated within the growing field of citizenship theories? In short: is environmental citizenship just another label, or one that signifies that something profound is happening to citizenship? Different theorists provide different answers to these questions (as we shall see in Chapter 3). But this text is not just about theories and theorists. The text is also about practices. In fact, the answer to those questions will probably lie in the actions of citizens, governments and corporations. If so, what do the actions of citizens, government policies and corporate initiatives tell us about the present and future of citizenship, specifically of environmental citizenship? Do they point to a green revolution, or to something else altogether (e.g. a neoliberal revolution)? In addition, the answer might also lie in part in the models of citizenship we teach our children (and subsequent generations). If so, what types of citizenship are being taught? What kinds of citizens are being manufactured by formal education systems and popular culture representations? These are all questions that are explored in this text. The task is structured as follows.

Structure of the book

The text is structured into three parts: Part I Concepts and Theories (Chapters 1–3), Part II Actions and Practices (Chapters 4–6), and Part III Pedagogies and Representations (Chapter 7). The content of the three parts is closely interrelated, reflecting the dynamics of the world outside the text. In that world, theories, actions and representations are in constant interaction. Theories explain practices or, in the case of normative theories, attempt to shape them. Practices impact on theories – at least to the extent that explanatory theories must incorporate them into their accounts or risk being discredited. Pedagogies are informed by theories and are themselves practices. And all of these interact with, and often become part of, the media and popular culture representations, which in turn inform and shape our practices and actions, and so the world goes round and round. If this sounds complicated it is because reality *is* complicated, and the reality of citizenship is definitely complicated. This alone justifies going back to basics, which is the point of Chapter 1.

Chapter 1 provides a brief introduction to the meaning, elements and concepts related to the two key themes and building blocks of the text: environment and citizenship. The content is often descriptive and serves to contextualize the emergence of the environment and citizenship in recent times as major issues and to provide a historical and conceptual scaffold to the two theoretical chapters that follow. The chapter reveals environment and citizenship as contested and relational concepts with long and fascinating histories. The environment is introduced as a term that refers to the relation between humans and nature, whilst citizenship is introduced as a term that refers to the relation between humans, between the individual and the collective. This chapter also provides an overview of the evolution of the modern environmental movement and a brief outline of the history of citizenship.

The next two chapters introduce the reader to theories of citizenship and explore the impact that environmental concerns are having on theoretical formulations of citizenship, both mainstream (liberal and republican) and alternative (pluralist and globalist). Chapter 2 builds on the outline of citizenship provided in Chapter 1. The chapter focuses on six theories of citizenship: two classical theories (liberal and republican), two pluralist theories (feminist and multicultural), and two globalist theories (cosmopolitan and neoliberal). The chapter serves as a map with which to navigate the complex and contested terrain of citizenship theories. This is a valuable exercise in and of itself, but is of particular relevance here; a necessary step to set out the basics for a full appreciation of the impact of the environment on contemporary theoretical formulations of citizenship, which is the task of Chapter 3. The chapter ends with the outline of neoliberal citizenship, signalling the fact that, of all the theories and models of citizenship, this one holds particular significance for what it means to be a citizen, environmental or otherwise, in the twenty-first century.

Chapter 3 explores how different theories of citizenship have been adjusted to integrate environmental concerns and ecological values. This chapter often mirrors the structure of the previous one, adding the environment to the theories of citizenship introduced in Chapter 2. In addition, the chapter introduces and explores novel theoretical articulations of environmental citizenship that have resulted from the incorporation of ecological values and concerns with sustainability into the theorizing of citizenship. In recognition of its widespread influence, the notion of 'ecological citizenship' theorized by Andrew Dobson is explored in a dedicated section of Chapter 3. This chapter concludes with an outline of alternative and emerging theories, approaches and reflections, many of which result from direct and explicit engagement with Dobson's work.

Part II explores the practical dimension of environmental citizenship, that is, the actions that shape and are informed by different conceptions of the relationship between citizenship and the environment. This part consists of three chapters, each focusing on a particular actor: citizens (Chapter 4), governments (Chapter 5), and corporations (Chapter 6). These chapters reflect the need to ground conceptual discussions of environmental citizenship in the concrete scenarios and actors where they operate, which is where the contents and meanings of citizenship become relevant (Jelin 2000: 48).

Chapter 4 explores the relation between conceptions of citizenship and the actions of environmental citizens, in particular, those of environmental activists. The chapter outlines the myriad ways in which environmental activism is shaping citizenship by greening existing conceptions of citizenship and by inspiring new articulations of environmental citizenship. The focus is on a selection of prominent movements, with attention paid to the different material and political contexts in which these operate, particularly those that differentiate between environmentalism in the Global North and the Global South. To that end, the chapter begins with a brief account of the framework that will be used here to classify environmental activism: 'the three posts' (i.e. postindustrialism, postmaterialism, and postcolonialism). The chapter illustrates how different contexts produce different articulations of environmental citizenship and points to the fact that some of the most comprehensive and original of these originate in the Global South.

Chapter 5 explores the greening of government with special attention to the main ways in which governments are deploying and shaping environmental citizenship in the process. The chapter begins by introducing the three main forms of governing modern societies (i.e. government, governance, and governmentality) and the neoliberal context in which 'the three govs' have operated in recent times, particularly since the 1990s. The chapter maps the incorporation of citizenship into the greening of government within a shifting historical context, from the top-down 'government approach' of the early decades, to the network-based 'governance approach' that has become the dominant paradigm since the 1990s, and the 'governmentality approach' that is gaining momentum in the twenty-first century. The different approaches are illustrated with examples from around the world, with particular attention paid to their articulation in the Global North and the Global South. The content reveals how environmental citizenship has become part of a more diffuse and often hidden form of governing societies and resolving environmental issues.

Chapter 6 explores the impact corporations have on the formulation and articulation of environmental citizenship, with attention to the main debates and controversies surrounding the concept of corporate environmental citizenship. The chapter begins by discussing the question of whether corporations are more like citizens (or quasi citizens) or more like governments (or quasi governments). The study illustrates how corporations have incorporated the language of environmental citizenship into their identities to the point that even the most consumption-oriented corporations claim to be model environmental citizens. The impact of this embrace is examined in relation to the three constitutive elements of citizenship: membership, rights and duties. The chapter concludes with a section on a novel conception of citizenship inspired by the operations of large retail corporations, the so-called supply-chain citizenship, exemplified here by Walmart.

The chapters in Part II include references to examples and case studies drawn from a wide range of countries, broadly categorized as belonging to the Global North and the Global South. This is an imperfect and contested distinction (as distinctions often are), but is also one that is worth keeping to differentiate between the affluent lives of the global minority, that is, those living in developed (and over-developed) countries (i.e. the Global North), and the less-affluent lives of the global majority, that is, those living in developing (or underdeveloped) countries (i.e. the Global South). The distinction also signifies the fact that the politics of globalization are still based on inherent inequalities and problematic power relationships between the Global North and the Global South. In matters environmental, the distinction also reflects the fact that, though we share the same planet and are affected by global trends, 'the same issues are different according to the perspective taken: from the South or from the North' (Jelin 2000: 47). Having said that, the increasing number, complexity and significance of global and glocal networks help to generate articulations of environmental citizenship that bring together agents and interests from the Global North and the Global South. In any case, and even if we take the distinction as a matter of emphasis, this should not lead us to dismiss its significance. Emphasis matters!

Chapter 7 explores the making of green citizens and outlines the dominant articulations of environmental citizenship that emerge from formal education, the news media and popular culture. The content of this chapter reflects the fact that citizens are made, not born. Citizens need some degree of knowledge, skills and dispositions, as well as capabilities, for their rights and duties to be meaningful. In other words, who is a citizen and what kind of a citizen one becomes are part of a highly manufactured process. The chapter identifies salient aspects and trends, both in the formal teaching of environmental citizenship and in media representations of green citizens. The content reveals the shift towards representations of citizenship that sideline the role of government, focusing instead on individuals and their interaction (as consumer-citizens) with corporations. The chapter pays special attention to the political socialization of children (the future citizens), and concludes with a section dedicated to exploring the most popular pedagogical tool used to promote environmental citizenship and manufacture green citizens: the ecological footprint.

Finally, the Conclusion brings together the main points raised in the text and concludes with some general reflections on what environmental citizenship offers to and demands from the citizens of the twenty-first century.

Preliminary considerations and additional materials

Textbooks require some level of selection and simplification. The fact that there is a lot of theoretical ground covered in this text means that the nuance and complexity of many theories (and the work of most theorists) are not fully acknowledged or explored. This is a necessary sacrifice in a textbook that seeks to develop the reader's understanding of the basic ideas, concepts and theories quickly and effectively. It is always important to bear this in mind, not least when encountering distinctions, categories and typologies. We should remain aware of the fact that categories are artificial constructions, albeit not arbitrary ones. This is an important consideration particularly when using 'the three posts' and 'the three govs' to navigate the content of Chapter 4 and Chapter 5, respectively.

The textbook contains a variety of illustrations, boxed content, additional resources, student activities and discussion questions. The selected readings that accompany each chapter are designed to encourage readers to further explore and engage with specific authors, concepts and themes. The material has been selected to reflect the diversity of content covered in each chapter, but also based on the significance and/or the engaging nature of the reading. In some cases, the reading is linked to an activity. The activities and questions are designed to encourage research and facilitate reflection and conversation, and can also be used to generate assignments (e.g. written essays) and structure classroom discussions. In addition, each chapter includes links to a variety of items (e.g. campaigns, organizations, documents, feature films and short videos). These links are invitations to further engage with the ideas introduced and discussed in the textbook. Most of the links relate to content covered in the text (in depth or in passing) and thus can be explored in conversation with the analysis presented in the book.

Part I
Concepts and theories

1 Environment and citizenship: the basics

> Nothing is more powerful than an idea whose time has come.
>
> (Victor Hugo)

Environmental citizenship is definitely an idea whose time has come. Governments around the world are recognizing environmental rights, activists refer to our duties to the environment, corporations present themselves as good environmental citizens, children are taught to be mindful of their ecological footprint, the media tell us repeatedly to reduce, reuse and recycle, and theorists have begun to consider all this under a new field of study called environmental citizenship. Throughout this book we will map and explore the varied and significant ways in which citizenship and the environment are converging, both in theory and practice. In doing so, we will see how citizenship is framing the environment in particular ways, and how environmental concerns are reshaping what it means to be a citizen in the twenty-first century. But before we launch into the exploration of the relationship between the two concepts, it is important to explore their meanings, histories and complexities. This will assist us when we come to reflect upon and evaluate the significance of their convergence.

Environment

Basic concepts

Environment is a contested concept. That is to say, there is no agreed definition, but rather a series of different understandings that shape any subsequent discussion about the environment. The roots of the term lie in the French word *environ*, meaning to surround, to envelop, to enclose. In this sense, environment is synonymous with surroundings.

But without a specific referent, this tells us little. We need to know what or who is surrounded in order to define an environment. In other words, environment is a relational concept. Thus, for example, Einstein reportedly said: 'The environment is everything that is not me' (Wilkinson 2002: 41). In this sense, the environment is the whole physical world that surrounds us. However, typically, the two elements of that relation are humans and nature, with nature taken as synonymous with the environment that surrounds humans, i.e. land, air and water, and the living organisms dependent upon these, namely, plants and animals, and microorganisms. This is the way in which the term will be used in this text. In short, the discussion about the meaning of environment is essentially a discussion about the relationship between humans and nature – a relationship shaped by how individuals and societies view and value the natural environment.

There are two main positions regarding the relationship between humans and nature: ecocentrism (nature-centred) and anthropocentrism (human-centred). Ecocentrism refers to the belief that nature has intrinsic value, regardless of its utility to humans. In some of its articulations, this approach grants equal value to all natural beings (plants, animals and humans). In essence, ecocentrism places the whole of nature at the centre of reality, and positions humans alongside all other living beings and natural elements. Anthropocentrism refers to the belief that the value of nature is relative to what it can provide for humanity. From this perspective, no inherent demand for environmental protection or nature conservation can exist beyond its potential to benefit humanity. In essence, 'anthropocentrism places humans at the centre of reality, and views humanity as standing apart from and above nature' (Postma 2006: 107). This belief argues that humans occupy a privileged position on Earth, and in nature, by virtue of our reason. Anthropocentrism can take two main forms: strong and weak. The first states that humanity can rightfully do with nature as it wishes. The second believes that humans have the responsibility to ensure that environmental practices are sustainable over the long term. This position states that since humans have advanced rational qualities, and even though we belong to the natural world (and are natural beings), we have a special place in nature that makes us particularly powerful and particularly responsible for the environment in which we (and all other organisms) live.

These different positions have long historical and philosophical traditions, which still inform how we view and act upon the natural environment. The dominant approach is anthropocentrism. The anthropocentric tradition has deep roots in Western culture and philosophy, and can be dated as far back as ancient Greece in the fourth century BCE. The most influential classical vision of the relationship between humans and nature is the work of the Greek philosopher Aristotle. His *Scala Naturae* (Great Chain of Being) positions humans above natural things, to which he attributed mere instrumental value (Figure 1.1). Aristotle believed that nature made all things specifically for the sake of man and accorded humans their privileged position above animals and plants, based on the notion that humans are the only living beings possessed of reason (*logos*).

The other major classical influence is the Bible. The holy text of Christianity places humans at the centre of creation and as masters of the natural world. In the Book of Genesis, God instructs Adam and Eve to replenish and subdue the Earth, to rule over the fish of the sea and the birds of the air, and over every living creature that moves upon Earth. This passage has often been interpreted as a licence to do as we please with the natural world; but others see the message of the Bible as one of 'stewardship' that does not condone the exploitation of nature (Barry 2007: 34–43). In this, the official view of the Catholic Church, man is entrusted to be the steward of God's creation. This interpretation often dwells on another passage from the Book of Genesis, the story of Noah's Ark. The distinction is crucial for it legitimizes a very different treatment of the planet and its creatures. The notion that humans are the 'vice-regents' under God with the responsibility to care for all creation is also present in Islam (Barry 2007: 33).

Figure 1.1 *Aristotle's* Scala Naturae, *or Great Chain of Being*

The anthropocentric tradition was strengthened during the centuries that encompassed the Scientific Revolution, the Industrial Revolution and the Enlightenment. This period produced what came to be known as 'the domination of nature' approach. The first and foremost exponent of this approach was the English philosopher Francis Bacon, who embraced the notion that nature was 'the gift of God' (Sarre and Brown 1996: 92). He stated that nature must be 'bound into service' and made a 'slave' to mankind – as humanity was typically referred to at the time. The philosophers of the Enlightenment subsequently developed a mechanical view of nature which conceived of reality as a machine comprised of discrete and individual parts whose actions could be known, and mastered for the benefit of humanity. This mechanical view of nature resulted in a profoundly utilitarian perspective that came to see the natural environment as a hostile domain to be controlled and exploited, with the help of science and technology, in the name of human progress (Barry 2007: 43–48). The domination of nature thesis only came to be seriously challenged in the second half of the twentieth century, following the emergence of the modern environmental movement. However, the anthropocentric view still prevails, albeit typically in its 'weak' manifestation.

The ecocentric view can be found in ancient paganism, indigenous cosmologies and some of the religious traditions of Asia (e.g. Buddhism, Hinduism and Shintoism). In general terms, these traditions see humans as part of the natural world, display a marked respect for the natural environment, and seek to promote a harmonious relationship between humans and nature. Buddhism, for example, holds that all forms of life (human and nonhuman) are interdependent and that we should avoid harm to other living beings (Barry 2007: 33). The desire to harmonize humans and nature can also be found amongst the English Romantic poets of the nineteenth century (e.g. William Blake, John Keats, Lord Byron), and amongst the Transcendentalist American literature of the nineteenth century (e.g. Ralph Waldo Emerson, Henry David Thoreau, Sarah Margaret Fuller), the high point of which is Henry David Thoreau's *Walden* (1854). In general terms, these authors held a view of nature as fragile, pure and pristine beauty, in need of protection. Trascendentalists displayed a preference for organic and dynamic conceptions of the natural world, and were particularly interested in the spiritual connections between humanity and nature with God. Their take on nature contrasts sharply with the rationalism of the Enlightenment. They rejected the scientific and religious narratives that presented humans as the masters of the natural world and embraced nature as the source of human enlightenment (Opie and Elliot 1996: 20–22).

The most influential ecocentric articulation of the twentieth century is that of American ecologist Aldo Leopold, particularly his book *A Sand County Almanac* (1949). Leopold developed a conception of the land as an ecological community to be loved and respected. His 'land ethic' was based on the notion that 'there is inherent worth in the integrity of natural ecosystems apart from any value they may possess for humans' (De Steiguer 2006: 15). His work, aimed at extending our moral concern to the natural environment and promoting harmonious relations between humans and the land, has left a profound impact on the conservation movement. The other major ecocentric figure of the last century is Norwegian philosopher Arne Naess. His work originated the tradition known as 'deep ecology', a term he coined in his seminal essay 'The Shallow and the Deep' (1973). Naess rejected the separation between humans and nature in favour of a relational image of nature based on 'biospherical egalitarianism' in which all ecological beings depend on each other and have an equal right to exist. His work provided a wholesale normative critique of modern human society, particularly of the modern human relationship with nonhuman nature, and inspired the emergence of ecocentric environmental groups in the twentieth century.

The way we conceive of the relationship between humans and nature shapes how we interact with the natural environment and the kinds of actions and policies we will support or reject when it comes to dealing with environmental issues – including, for example, the level at which we might include or exclude beings other than humans (e.g. nonhuman animals) in our conceptions of citizenship. In this regard, Andrew Dobson has theorized an important distinction between two forms of green political thought that inform two kinds of green politics, derived from the different understanding of the relationship between humans and nature: ecologism (or dark green politics); and environmentalism (or light green politics). The dominant form of green politics, light green politics, is informed by environmentalism, which 'argues for a managerial approach to environmental problems, secure in the belief that they can be solved without fundamental changes in present values or patterns of production and consumption' (Dobson 2007: 2). This approach is underpinned by a sense of human control of the nonhuman natural world that derives from the modernist, anthropocentric notion of the 'domination of nature' and is reflected in contemporary notions of 'sustainable development' and 'ecological modernization'. In contrast, the less common (and more radical) form of green politics, dark green politics, derives from ecologism, which 'holds that a sustainable and fulfilling existence presupposes radical changes in our relationship with the non-human natural world, and in our mode of social and political life' (Dobson 2007: 3). This approach is underpinned by a sense of human frailty and ultimate dependence on the environment, as well as by an appreciation of nature for its own sake, and not simply or necessarily in relation to human life. Instead of a managerial approach, ecologism takes an ethical approach that emphasizes care and justice, rather than control of the nonhuman natural world.

Historical overview

The environment emerged as a major point of contention and an issue of public and political concern in the 1960s. The publication of Rachel Carson's *Silent Spring* (1962) was a key moment. *Silent Spring* was a lucid and emotive account of the dangers of the systematic use of pesticides, in particular, dichlorodiphenyltrichloroethane, commonly known as DDT. Carson made use of the mass media to great effect, in the hope that by transmitting awareness of the environmental destruction caused by pesticides, the action of democratic citizens could change

Figure 1.2 *Rachel Carson, author of* Silent Spring *(1962)*

Source: Wikimedia Commons.
Author: U.S. Fish and Wildlife Service.

government policies for the benefit of the environment and in the interest of public health. Other critical works published in this period also heightened environmental awareness and led to the development of environmental sciences as an academic discipline. Some of the most influential were Gareth Hardin's 'Tragedy of the Commons' (1968) and Paul Ehrlich's *The Population Bomb* (1968). Hardin's essay examined how individual rational behaviour over the use of the natural commons could lead to collectively irrational outcomes, whilst Ehrlich's book argued that population growth was an 'explosion' with the potential to destroy the planet's ecological balance.

These early concerns with pollution and public health led citizens to mobilize, with protest movements incorporating environmental concerns and spawning the first environmental movements and the founding of some of the best-known and most influential environmental non-governmental organizations (NGOs), such as the World Wide Fund for Nature (WWF), founded in Morges (Switzerland) in 1961, Friends of the Earth (FoE), founded in San Francisco (the USA) in 1969, and Greenpeace, founded in Vancouver (Canada) in 1971. These groups had some early success in influencing government policy, particularly in the United States, where a raft of environmental legislation was passed in the 1960s, including the Clean Air Act of 1963, the Solid Waste Disposal Act of 1965, and the Wilderness Act of 1964. The trend continued in the 1970s with the creation of the Environmental Protection Agency in 1970, and the passing of the Clean Water Act of 1972 and the Endangered Species Act of 1973. These Acts were revised several times in subsequent years, strengthening the protection of the environment and human health with each amendment.

The surge in environmental awareness that dominated the 1960s was followed by an overwhelming feeling of environmental crisis in the 1970s. This feeling was partly informed by the alarming projections that came out of a study on 'the predicament of mankind' commissioned by the Club of Rome, a global think tank founded in Rome in 1968. The report, entitled *The Limits to Growth* (1972), argued that in a world of finite resources and ever-increasing demands on such resources, driven by increasing consumption and growing population, there was a need for radical changes to avert a global catastrophe in the second half of the twenty-first century (Meadows *et al*. 1972). This discourse seemed to be vindicated by the energy crisis of 1973, precipitated by the Yom Kippur War between Israel and the Arab countries. The publication of the report coincided with the United Nations Stockholm Conference on the Human Environment (1972), the first global conference to deal with environmental issues comprehensively and led to the creation of the United Nations Environment Programme (UNEP). The transnational nature of many environmental threats made it quite clear that geopolitical and economic change was necessary.

The 1980s also signalled the origins of ecological politics. Ecopolitics led to the birth of some of the most radical environmental groups, most notably Earth First!, founded in 1980 and inspired by Rachel Carson's *Silent Spring*, Aldo Leopold's land ethic and the tactics outlined in Edward Abbey's *The Monkey Wrench Gang* (1975). Their activism reflected the direct action approach espoused by Abbey, and their politics were neatly captured in their official slogan: 'No compromise in defense of Mother Earth' (Cudworth 2003: 95). These groups were also influenced by the concept of 'deep ecology' formulated by Naess, and his argument that our environmental predicament was not a 'shallow' problem of technical mastery, but a 'deeper' problem of mentality and worldview that required a reinterpretation of our relationship with nature and a change in our basic attitudes towards the natural environment. His analysis of the environmental crisis and his ethics of interrelatedness, based on the notion that all forms of life are equally entitled to live and flourish, continue to influence ecological politics into the twenty-first century (Taylor 2010).

The sense of environmental crisis accelerated in the early 1980s, with increasing public and media concern over issues such as global warming, acid rain, the depletion of the ozone layer and tropical deforestation. This decade signalled also the definitive explosion of environmental issues into the public imagination and the proliferation of environmental groups, organizations and movements. The growing interest in the environment was partly aided by a series of catastrophic accidents and tropical wildfires that made headlines around the world (Box 1.1). These disasters highlighted that there were reasons for serious and global concern regarding the human impact on the environment. During this decade the nuclear issue also emerged as central to the environmental movement, particularly in the Global North. The preoccupation with nuclear energy gained prominence after the accident at Three Mile Island, in the United States in 1979, and intensified following the Chernobyl nuclear meltdown in the Ukraine in 1986. In West Germany, whose population was sensitive to the anti-nuclear message, that preoccupation translated into the creation of the Green Party (Die Grünen) in 1980. Nuclear energy has remained a significant point of contention in environmental politics

Box 1.1

Major environmental disasters

- *The Three Mile Island nuclear accident, Pennsylvania, USA (1979)*: When some water pumps failed, there was a partial meltdown of one reactor core resulting in the release of radiation. Whilst there were no victims, the incident caused a major public safety alert. This was the first significant nuclear accident and is still the most serious in US history.

- *The Bhopal chemical disaster, India (1984)*: A gas leak in a chemical plant run by the American company Union Carbide caused the immediate death of over 2,000 local residents of Bhopal, and hundreds of thousands of appalling physical and psychological injuries. The total death toll remains disputed, ranging from the official figure of 3,787 to well over 20,000 disaster-related deaths. The surviving victims continue to fight for appropriate compensation and justice. This is still the largest industrial disaster in world history.

- *Amazon jungle fires, Brazil (early 1980s)*: Massive fires broke out across the Brazilian jungles in the early 1980s as a result of slash and burn rancher activities to clear land for farming. The extent of the fires attracted international attention to ruinous farming practices and to the need to protect large tropical forests, especially the Amazon.

- *The Chernobyl nuclear explosion, Ukraine (1986)*: The explosion of a nuclear reactor sent a cloud of radioactive material into the atmosphere that spread across Europe. The health effects of this meltdown have been highly contested since then, but the incident is symbolic of the worst possible outcomes associated with nuclear energy. This was the world's worst nuclear accident, until the Fukushima disaster of 2011.

- *The Exxon Valdez oil spill, Alaska, USA (1989)*: An oil tanker, the *Exxon Valdez*, spilled over 11 million gallons (41.8 million litres) of crude oil into an extremely environmentally sensitive area of Alaska. This was the largest oil spill ever to take place in the US until the Deep Water Horizon accident of 2010.

- *Deep Water Horizon or BP oil spill, Gulf of Mexico, USA (2010)*: A wellhead blowout at a deep water ocean oil rig caused a catastrophic oil spill, killed 11 workers and caused a fire that burned for 36 hours before the rig sank, ironically, on 22 April, Earth Day. As engineers scrambled to find a solution, the well continued to gush oil, until it was finally capped on 15 July. There is no exact estimate of how much oil was actually spilled, but the US government and BP have spent billions cleaning up the mess.

- *Fukushima Daiichi nuclear disaster, Japan (2011)*: Catastrophic failure of the Fukushima nuclear plant after fuel rods were exposed following the tsunami of 2011. Although no short-term deaths from radiation were reported, thousands of people were evacuated and the operator, Tokyo Electric Power Co (Tepco), continues to struggle to decommission the plant, which is now allegedly leaking radioactive water into the Pacific Ocean.

ever since and regained prominence following the nuclear meltdown in Fukushima, Japan, in 2011.

The most significant development of the 1980s was the integration of environment and development into the global political agenda that followed the publication of *Our Common Future* (1987). Better known as the Brundtland Report, this text was the work of the World Commission on Environment and Development (WCED), established by the United Nations in 1983. The commission was tasked with articulating a long-term vision for the world that would bridge the gap between environment and development, and between the developed and developing countries. Their report echoed the increasing realization that poverty and pollution were intimately related – a sentiment captured in the words of then Indian Prime Minister, Indira Gandhi: 'Poverty is the greatest polluter' (Cederlöf and Rangarajan 2009: 222). The Brundtland Report positioned sustainable development at the core of the global environmental agenda, defining it as: 'development that meets the needs of the present without compromising the ability of future generations to meet their own needs' (WCED 1987: 12).

The concept of sustainable development was institutionalized at the 1992 UN Conference of Environment and Development (UNCED) in Rio de Janeiro in Brazil, popularly known as the Earth Summit. The conference deployed the language of sustainable development to advance an international movement to manage the global environment, linking environmental degradation with poverty and over-consumption. The summit also produced a number of important international documents, most notably the Rio Declaration on Environment and Development and Agenda 21, the action plan for sustainable development for the twenty-first century. The institutionalization of sustainable development within these documents and others that followed 'is probably the most significant shift in environmental policy ever witnessed' (Barr 2008: 29). Since then, sustainable development has come to represent the mainstream thinking about the relationship between environment and development, and has been widely endorsed by governments, institutions and other organizations, both in the Global North and the Global South.

The widespread appeal of sustainable development comes from the fact that it tries to reconcile the ongoing aspirations of the poor to improve their living conditions with concerns about the environmental damage that often results from economic development. In other words, sustainable development tries to balance the concern for the environment with the needs of those who are neglected by present economic and political structures, namely: people living in underdeveloped countries and future generations (Postma 2006: 6). Yet, the concept has a clear anthropocentric nature, best reflected in the first of the 27 principles of the Rio Declaration: 'Human beings are at the center of concerns for sustainable development' (UNEP 1992a).

The term sustainable development has been broadly accepted, not least due to its lack of specificity. The general consensus around sustainable development as the mainstream framework for global environmental policy rests upon the vague substance of the term sustainability, one that leaves much room for interpretation. Indeed, there are hundreds of documented definitions of sustainability and sustainable development (Keiner 2006: 1–3). This is not to say these terms lack meaning or substance. However, in the absence of a precise meaning, the struggle between those who prioritize development and those who prioritize sustainability will continue unabated. In any case, the widespread adoption of the term sustainable development has placed the environment at the heart of mainstream global politics.

The turn of the century opened also what Gregg Easterbrook called 'The Coming Age of Environmental Optimism' – the subtitle of his book, *A Moment on the Earth* (1995). Easterbrook argues that the worst is nearly over. Technology has finished a phase of creating environmental problems and is now entering a phase of solving them (Easterbrook 1995: 267). His self-proclaimed 'ecorealism' echoes the optimism of earlier critics of the 'limits to growth' approach, exemplified most notably by Julian Simon and Herman Kahn's edited volume, *The Resourceful Earth* (1984), and Bjørn Lomborg's *The Skeptical Environmentalist* (2001). The optimism of these authors is anchored in what David Pepper calls 'technocentrism': a mode of thought that recognizes environmental problems but believes that humans always will be able to solve them and achieve unlimited growth (cornucopians) or that we will be able to negotiate them with careful economic and environmental management (accommodators) (1996: 37).

Technocentrism is at the heart of ecological modernization, the paradigm of ecological economics that has come to dominate environmental politics (and policy) in the early twenty-first century. First proposed by Joseph Huber in the 1980s, the essence of ecological modernization is the notion that economic growth can be decoupled from environmental harm through 'the invention, innovation and diffusion of new technologies and techniques of operating industrial processes' (Murphy 2000: 3). Ecological modernization implies 'a partnership in which governments, businesses, moderate environmentalists, and scientists co-operate in the restructuring of the capitalist political economy' (Dryzek 1997: 144). The reform of the capitalist system envisaged by ecological modernists does not challenge the capitalist mode of production and the consumer culture of late modernity. Ecological modernization has become a central concept in the institutional response to environmental challenges, to the point that some refer to it as 'a new, and in many ways improved, synonym for sustainable development' (Buttel 2000: 63). Its critics argue that whilst the concept may offer some hope for better environmental outcomes, it can equally serve as 'a cover for business-as-usual with a slight green tinge' (Gibbs 2000: 15). In fact, it is perfectly reasonable to assume that ecological modernization has become so popular because it suggests that we can have our cake and eat it too.

Sustainable development and ecological modernization are currently intertwined with debates about climate change. This issue has become prominent in the global environmental agenda since the 1980s, partly thanks to the regular assessment reports released by the Intergovernmental Panel on Climate Change (IPCC), the scientific intergovernmental body established in 1988 by two United Nations organizations, the World Meteorological Organization (WMO) and the UNEP. Their reports have received widespread media coverage and the organization was awarded the Nobel Peace Prize in 2007, alongside former US Vice-President Al Gore. However, in the latest twist to the ongoing story of environmental politics, the sense of urgency regarding climate change generated around the turn of the century has been dampened by the impact of the global financial crisis (Figure 1.3).

The global financial crisis has placed economic growth back at the centre of the global political agenda, even though some notable economists, such as Lord Stern (in the UK) and Ross Garnaut (in Australia) have produced substantive reports making a strong economic case for the urgent and decisive tackling of climate change. In the absence of a global agreement on climate change, which is partly due to the inability to settle the 'climate debt' between the Global North and the Global South, many continue to advocate technological solutions, such as geo-engineering, in line with the paradigm of ecological modernization (Goodell 2010; Kintisch 2010). Industrial activities and technologies always carry significant risks for human health and the natural environment, most recently evidenced by the oil spill in the Gulf of Mexico and the nuclear

Figure 1.3 *'The Sky is Falling!'*

Source: Cartoon courtesy of Cathy Wilcox.

disaster in Fukushima. The fact that proposals to use technology to alter the planet's climate continue to be contemplated, despite these recent catastrophes, betrays an almost delusional faith in human ingenuity and/ or a significant level of desperation regarding current climate change trends.

In the meantime, the global environmental and developmental agenda has shifted even more towards the economic, illustrated in the adoption of the concept of 'green economy' at the 2012 Rio+20 UN Conference on Sustainable Development. The logic of ecological modernization is perfectly reflected in the words of UN Secretary-General Ban Ki-moon, in a statement issued on the release of the UNEP's flagship report, 'Towards a Green Economy: Pathways to Sustainable Development and Poverty Eradication' (2011). He writes: 'With smart public policies, governments can grow their economies, generate decent employment and accelerate social progress in a way that keeps humanity's ecological footprint within the planet's carrying capacity' (cited in UNEP 2012b). In contrast to this optimism, critics have called the 'green economy' a 'red herring' that fails to confront 'the basic contradiction between ever-expanding human activity and a finite world' (Spash 2012a: 98).

The extent to which these fundamental tensions or contradictions will be confronted or avoided, resolved or endured, will be partly related to the role played by another important approach to environmental politics that is becoming increasingly influential: environmental citizenship. This emerging approach, as we already know, is the focus of this text. So, having considered its first pillar, the environment, it is now time to turn our attention to the second: citizenship.

Citizenship

Basic concepts

Citizenship is one of the oldest social and political institutions in human history and one of the fundamental systems that structures human relations in modern societies. Like environment, citizenship is a notoriously polyvalent and contested concept. This complexity derives in part from the fact that citizenship has a long history that goes back to ancient Greece and Rome. Indeed, the word citizen comes from the word *civis*, the Latin version of the Greek word *polites*. The Greek *polites* and the Roman *civis* were terms that referred to the members of the Greek *polis* (city-state) and the Roman *civitas* (city-state), respectively. The etymology of citizenship is also related to the use of the word in medieval times. The medieval citizen was above all an economic citizen, the burgher, whose civil rights were connected with the protection of private property and commercial transactions. It was in reference to the inhabitants of the burghs/cities (the bourgeoisie) that the term citizen first came to be commonly used in English (Smith 2002: 107). With the emergence of the modern nation-state in the seventeenth century, citizenship came to be associated with nationality – the legal expression of membership of a nation-state. However, the simplicity of this definition belies the widespread debates regarding the meaning, dimensions and applications of citizenship (Joppke 2007).

The complex and contested nature of citizenship derives also from the fact that citizenship is a relational concept. Citizenship is a relationship between the individual and the collective, between the citizens and the political community to which they belong. The relational nature of citizenship means that 'the content of enforceable rights and obligations that tie together states and citizens are historically and geographically variable' (Barnett and Scott 2007: 294). In other words, citizenship is always and everywhere in a permanent process of construction and transformation. However, for all its diversity and dynamism, there are

three elements that tend to be present in conventional definitions of
modern citizenship: membership, rights and duties. These elements
refer to the three essential dimensions of citizenship: the affiliative
(affiliation); the legal (protection); and the civic (participation). Thus,
before we embark upon the examination of citizenship theories (the
task of Chapter 2), it is important to get a sense of these elements,
beginning with the one that is arguably the most complex of the three:
membership.

Membership

Membership lies at the heart of citizenship. In fact, citizenship is
sometimes defined simply as membership of a political community.
In modern times, this means membership of a nation-state. Membership
is about the 'who' of citizenship, that is, about determining who is a
citizen. In the modern sense, citizenship 'is about who belongs to the
nation, who does not, and why' (Gorman 2006: 1). The answers to
these questions come, first and foremost, from the formal rules that
each political community has in place to grant or deny citizenship.
The specific rules are complex, vary significantly across countries
and are in a permanent state of flux. These rules refer to the concept
of citizenship as an 'allocative institution' (Brubaker 1992: 26). From
this perspective, citizenship can be defined as 'an international filing
system, a mechanism for allocating persons to states' (Brubaker 1992:
31). However, citizenship as membership is far more complex and
meaningful than the mere allocation of people to nation-states.
Membership refers also to the sense of belonging and identity associated
with being part of the political community. Indeed, citizenship is
'a primary means through which societies assert, construct, and
consecrate their sense of identity' (Gorman 2006: 1). In this sense,
membership relates to the affiliative dimension of citizenship, that is, to
the traits and sentiments associated with the personal affiliation with the
political community.

Membership reveals one of the fundamental dynamics of citizenship:
inclusion vs exclusion. The dynamic operates at two different levels:
internally (within the political community) and externally (between
political communities). On the one hand, citizenship draws social
boundaries between different groups of people living inside the
community, i.e. between groups with access and groups without access
to the public sphere. On the other hand, citizenship draws political

boundaries between different communities, i.e. between insiders and foreigners. The tensions created by this double dynamic have informed much of the evolution of citizenship. Internally, the history of citizenship is a history of the challenges posed to each and every one of the traditional qualifications for membership (i.e. class, property, gender and ethnicity), present since the birth of citizenship in ancient Greece. Externally, the history of citizenship is a history of the redrawing of boundaries between political communities. In short, the history of citizenship as membership of a political community is a history of two trends: internal inclusion (though this process is far from complete) and external exclusion (still a fundamental element of the current dominant model of citizenship, namely national citizenship).

Membership creates two basic types of individuals: citizens and non-citizens. In modern times, this translates into nationals and foreigners (or aliens). However, this distinction does not provide the full picture of citizenship as membership. There are numerous examples of incomplete membership, the most prominent of which are second-class citizens, marginal citizens (or margizens) and denizens. The first term refers to individuals who have the legal status of citizens but because of discrimination are denied full rights (e.g. women in some countries, sexual minorities in most). The second term refers to people who have the legal status of citizens and are formally granted full rights but whose level of social, economic and cultural capital is so poor that they are effectively excluded from citizenship (e.g. the homeless, the ultrapoor). These citizens often enjoy less substantive citizenship than denizens (e.g. foreign residents who, because they are not citizens, lack or have limited political rights, but enjoy civil, social and economic rights). Having said that, non-citizens often fill the ranks of the marginalized. These include foreign residents who lack secure residence status (e.g. illegal workers), unauthorized family entrants, asylum-seekers, 'permanent' temporary workers, unemployed migrants, etc. These intermediate categories reveal another fundamental dynamic of citizenship: the tension between formal citizenship and substantive citizenship. In other words, they show that formal membership does not equate with the ability and capacity to enjoy the rights and exercise the duties associated with citizenship (Castles and Davidson 2000).

Membership also highlights the relational character of citizenship. This aspect is perfectly captured in the definition of modern citizenship provided by the *Encyclopedia Britannica*: 'Citizenship is a relationship between an individual and a state in which an individual owes allegiance to that state and is in turn entitled to its protection' (cited in Frey 2003: 95). This contractual arrangement that ties the individual and the state has been complicated in recent times with the increasing proliferation of dual citizenship, nested citizenship, and other forms of multiple citizenship that challenge the idea of a single loyalty (Sejersen 2008). The relationship between the individual and the state has been further complicated by the incorporation of a third actor: the corporation. The inclusion of corporations has given additional force to the contractual and transactional character of modern citizenship. In any case, the notion of citizenship as a series of transactions refers to the content of citizenship: rights and duties.

Rights

Citizenship is often identified with rights. Hannah Arendt famously defined citizenship as the ultimate right: 'the right to have rights'. Rights are about the 'what' of citizenship, that is, about its content. The specific rights associated with citizenship (their type, number and relative value) have varied widely across time and space. However, and at the risk of oversimplifying the historical portrait of citizenship rights, we can say that political rights (albeit treated simultaneously as duties) were the essence of citizenship in ancient Greece, civil rights (at the expense of political rights) became the trademark of citizenship in imperial Rome, and commercial rights were the hallmark of urban citizenship in medieval Europe. The modern portrait of rights originates from the work of English sociologist T.H. Marshall, particularly his seminal essay 'Citizenship and Social Class' (1950).

Marshall distinguished three types of citizenship rights: civil rights, political rights and social rights. Civil rights are rights necessary for individual freedom, including rights that protect the individual against the actions of the state, such as freedom of expression, freedom of religion, the right to own property and conclude valid contracts, freedom from arbitrary detention, the right to a fair trial and equality before the law. Political rights are rights that enable citizens to participate in the political process, such as the right to vote and the right to stand for office, but also freedom of assembly and association, and freedom of information. Social rights are entitlements to social and economic provisions designed to guarantee a decent standard of living, and include the right to work, the right to education, the right to health and the right to a basic income or pension for those out of work or unable to work.

This typology of rights is still the most influential. However, it is important to keep in mind that the three categories of rights are not rigid or mutually exclusive: e.g. free speech can be regarded both as a civil and a political right; and the right to property (often listed as a civil right) can be considered an economic right. It is also important to realize that the different rights are interconnected and ultimately interdependent: e.g. civil rights are essential to exercise political rights; and political rights are essential to protect civil rights. Marshall's analysis showed how civil and political rights were crucial to establish social rights, but also how social rights are essential to fully enjoy civil and political rights. The interconnection and interdependence of rights have remained central to our understanding of citizenship ever since. In recent times, others have argued the need for additional rights to guarantee equal and full citizenship for a range of different social groups, including women, sexual minorities, ethnic minorities and indigenous groups (e.g. Richardson 1988; Kymlicka 1995).

The expansion of rights has been the main trend in relation to citizenship, especially in the second half of the twentieth century. This has led some to view citizenship as an expansive sphere in which new rights are added to a growing body of rights as new social forces are included in the national community (Barbalet 1988). However, we should not take the expansion of rights as an inexorably unfolding process driven by the march of progress, but instead we must view all rights as sites of contestation. In recent times, for example, we have seen increased pressure to unwind social rights associated with the welfare state and a significant backlash against the cultural rights of ethnic minorities. In this sense, citizenship can also be defined as 'a social process through which individuals and social groups engage in claiming, expanding, or losing rights' (Isin and Turner 2002: 4). The constant struggle over rights indicates that we must resist linear narratives of citizenship. Instead, that struggle suggests that rather than assume the battle for particular rights has been won, we must appreciate the contingent and contextual nature of the process through which citizenship develops. We can also appreciate this through the ebb and flow of the significance attached to rights relative to the third and final element of citizenship explored here: duties.

Duties

Duties have been integral to the concept of citizenship since its inception. Like rights, duties are also about the 'what' of citizenship, that is, about its content. However, the picture of duties is less complicated than that of rights. The most common duties have been: obeying the law, paying taxes, and some form of public service (e.g. jury duty and military service). These duties have been rather constant throughout history – the main exception being military service, which has seen a considerable reduction in recent times, often replaced with some form of social service (e.g. elderly care, school help). The main change when it comes to duties is not their number or type but their weight relative to that of rights. In general terms, duties prevailed in ancient Greece, rights prevailed in ancient Rome, both were limited in medieval Europe, and rights have prevailed over duties in modern times, particularly since the twentieth century. This constant rebalancing of rights and duties is another reminder that there is no necessary historical logic underpinning the evolution of citizenship (Turner 1986).

The tension between rights and duties is at the heart of citizenship, but we must not assume (as is often done) that rights and duties are always in competition, let alone in direct and almost irreconcilable contraposition. Instead, we must keep in mind that the two are intimately related. On the one hand, rights facilitate the performance of duties. Thus, for example, social rights enable citizens to participate in the economic life of the community, which in turn allows them to pay taxes. On the other hand, rights depend upon the effective discharge of duties. Thus, for example, the protection of civil rights and the provision of social rights cannot proceed without at the very least the payment of taxes. Similarly, the exercise of political rights cannot proceed without the discharge of political duties, which is why in some countries (e.g. Australia) voting is compulsory, that is, both a political right and a formal civic duty. In other words, in the same way that we should not lose sight of the interdependence between different rights, we must also keep in mind the interdependence between rights and duties.

The precise rights and duties, the balance between the two, the relative importance attributed to specific rights and duties, and the relative emphasis placed on rights and duties as a whole have varied significantly over the centuries and across communities, but rights coexist with duties in all conceptions of citizenship. In the end, the relative weight and significance given to each are a matter of emphasis, but emphasis matters. Indeed, the different emphasis given to rights and duties is one of the key aspects that differentiates the two classical and still mainstream theories of citizenship: liberal and republican. The former defines citizenship as a status, and is fundamentally a rights-based approach, which prioritizes legal protection and individual freedom. The latter defines citizenship as action, and is fundamentally a duties-based approach, which prioritizes political participation and the common good. These two traditions have deep historical roots, as has citizenship in general. This makes it imperative to approach citizenship with 'historical sensibility' (Dobson 2003: 35). So, before we go any further, let us take a brief journey through the history of citizenship, from its ancient origins to where it is at in the twenty-first century.

Historical overview

The origins of citizenship are commonly associated with the birth of democracy in ancient Athens, some time between the sixth and fourth century BCE. Athenian citizenship was synonymous with participation in the everyday running of the community (the *polis*, or city-state). The true citizen, or at least the good citizen, was the active citizen who devoted his life to public affairs. Those who preferred a more private or less arduous life could find themselves mocked as 'good for nothing'. But we must be careful not to automatically associate citizenship with democracy. Throughout history, citizenship has been an 'ambiguous institution' and has been compatible with different forms of political organization (Riesenberg 1992). In fact, the other major Greek influence in the origins of citizenship, Sparta, was a city-state with a similar conception of citizenship to Athens, but in the context of an authoritarian regime and a militaristic culture that produced the ideal 'citizen-soldier'. In addition, citizenship was a highly exclusive club, restricted to free, adult men of Spartan and Athenian descent, respectively.

Although inspired in ancient Greece, citizenship took on a new and different meaning in ancient Rome. Roman citizenship came to be associated with legal protection rather than political participation. It did not imply active citizens who exercised political power, but passive citizens under the legal protection of the Roman Empire. Moreover, citizenship was not exclusive to ethnic Romans, but was extended to new peoples as the power of Rome expanded, first across the Italian peninsula and then across Europe and into Asia and Africa. Thus, citizenship during the Roman Empire represented a step forward in terms of inclusiveness, but a step backwards in terms of participation (and democracy) when compared with citizenship in ancient Greece (Castles and Davidson 2000: 33).

The collapse of the Roman Empire left individuals at the mercy of local authorities, with no formal recourse to the legal protection of Rome. Citizenship did not vanish entirely, but was eclipsed by the various feudal and religious statuses of the medieval Christian world. The classical tradition of citizenship that required loyalty to a single political authority was replaced by personal, multiple and often conflicting loyalties to secular and religious authorities (e.g. lords, monarchs, bishops and popes). In medieval times, authority was 'personalized and parcelized within and across territorial formations' and there was no 'notion of firm boundary lines between the major territorial formations' (Ruggie 1998: 179). This picture of fragmentation and confusion had only one relatively minor but important exception: the cities. Citizenship came to signify 'the relationship of freely exercised rights and duties in a city or town' (Heater 2004b: 22). The most sophisticated and complete expression of medieval citizenship was found in the city-states of Italy, particularly Florence, Milan and Venice, the embryonic political communities of the modern nation-state in Europe. Importantly, whilst the practice of citizenship had been severely affected during the Middle Ages, the ideas of citizenship remained relatively unchanged from the classical world through to the eighteenth century.

The transition from medieval to modern citizenship was a long and arduous process driven largely by the growth of capitalism and the revolutionary changes that took Europe out of the Middle Ages and into the Modern Era. Gradually, territorial borders began to consolidate and ruling became a more impersonal domain, done at a distance (from the royal court, or national capital), and distinct from family ties, religious affiliations and commercial relations. States became entities unto themselves, independent, supreme, sovereign; and citizenship became the exclusive membership of (and loyalty to) this new political community: the nation-state. The radical break with medieval citizenship came with the articulation of a single loyalty, not to a person (the prince or monarch) but to an abstract entity (the state). Initially, monarchs claimed absolute and untrammelled power over their subjects, defining themselves as synonymous with the sovereign state – a sentiment perfectly captured in Louis XIV's proclamation: 'I am the State'. The popular reaction against the absolute and arbitrary power of the monarch would come to define the origins of modern citizenship. The first and most successful challenge to monarchical power was the English Civil War, which led to the execution of Charles I in 1649. The same spirit of resistance to tyrannical rule informed the American Revolution (1776),

Figure 1.4 *The Declaration of the Rights of Man and the Citizen (1789)*

Source: Wikimedia Commons. Author: Jean-Jacques-François Le Barbier.

which led to the independence of the Thirteen Colonies from Britain
and the subsequent formation of the United States of America. The
culmination of the revolutionary spirit came with the French Revolution
(1789), which abolished the monarchy, declared the political supremacy

of the citizens (i.e. popular sovereignty), and formalized the modern contractual concept of citizenship in the Declaration of the Rights of Man and the Citizen (1789).

The modern conception of citizenship is based on the notion of the social contract, a metaphorical contract between the people (constituted as citizens) and the state (constituted as political authorities) by which citizens are afforded certain rights and protections in exchange for loyalty and certain obligations. The social contract typically takes the form of a constitution, the political document (typically agreed by the citizenry) that formalizes the specific rights and duties of both parties. The concept takes on slightly but significantly different meanings in the modern republican and liberal variations, in line with their classical roots in ancient Greece and ancient Rome. Modern liberal citizenship, best illustrated by the American Revolution and the work of John Locke, emphasizes legal protection and the rights of the individual. By contrast, modern republican citizenship, best illustrated by the French Revolution and the work of Jean-Jacques Rousseau, emphasizes political participation and the duties towards the community (Bellamy 2008).

The modern conception of citizenship extended across the world during the nineteenth and twentieth centuries. Since then the nation-state has become the dominant political community and national and contractual citizenship the dominant articulation of the relationship between the individual and the state. Nowadays the vast majority of people around the world see themselves as belonging to a territorially bound political community in which they have certain rights and duties, even if those rights are not always protected in practice. However, the nation-state is currently facing important challenges, arising in particular from the process of globalization, and this has generated significant debate about the prospects for national citizenship in the twenty-first century (e.g. Brodie 2004).

The fact that the history of modern citizenship is inseparable from the history of the nation-state has led some to argue that the two will inevitably share the same fate, whatever that might be. However, we must remember that national citizenship is only one articulation of citizenship, albeit the dominant one for the past two centuries. Thus, whilst the eventual demise of the nation-state would probably lead to the demise of national citizenship, this must not be confused (as it often is) with the demise of citizenship altogether. There are equally valid reasons to argue that the decline of the nation-state could bring about the end of citizenship as there are to argue that non-national versions of citizenship can emerge and prevail in the future. The outcome might not be determined in our lifetime, but there is a good chance that substantial changes to citizenship, as we know it, will occur during this, the twenty-first century. Some of those changes could be triggered by the increasing significance of environmental concerns. But before we can fully appreciate the impact the environment might have on citizenship, we need a more sophisticated understanding of citizenship theories. Providing a map to navigate these theories is the task of Chapter 2.

Summary points

- Environment and citizenship are both contested and relational concepts with long historical trajectories.
- The discussion about the meaning of the term environment is essentially a discussion about the relation between humans and nature. The two main positions regarding this relation are ecocentrism (nature-centred) and anthropocentrism (human-centred). These positions have long historical roots and inform contemporary environmental politics.
- The environment emerged as a major point of contention and a matter of public and political concern in the 1960s. Since then, engagement with environmental issues has taken several forms: awareness, crisis (e.g. limits and disasters), sustainable development, ecological modernization and, in recent times, environmental citizenship.
- The discussion about the meaning of the term citizenship is essentially a discussion about the relation between humans, particularly between the individual and the collective or the individual and the state.
- Citizenship has a long history, and has taken different shapes at different times and in different places. Yet, for all its diversity and dynamism, there are three elements that tend to be present in conventional definitions of citizenship: membership, rights and duties.
- Membership is about the 'who' of citizenship and reveals its relational character as well as one of its fundamental dynamics, namely, inclusion vs exclusion.
- Rights and duties are about the 'what' of citizenship. The specific rights and duties, and their relative importance, have varied historically, but they are both present in all conceptions of citizenship, and they are always interrelated and interdependent.
- The history of citizenship is long and in constant flux, and suggests that whilst national citizenship seems natural, different forms of

citizenship could emerge and prevail in the future – some might even come about as a result of increasing environmental concerns.

Selected readings

Bellamy, R. (2008) *Citizenship: A Very Short Introduction*. Oxford: Oxford University Press.

Brodie, J. (2004) 'Introduction: Globalization and Citizenship beyond the National State'. *Citizenship Studies*, 8(4): 323–332.

Carson, R. (1962) *Silent Spring*. Boston: Houghton Mifflin.

Carter, N. (2007) *The Politics of the Environment: Ideas, Activism, Policy*. 2nd edn. New York: Cambridge University Press.

Dryzek, J. S. (2005) *The Politics of the Earth: Environmental Discourses*. 2nd edn. New York: Oxford University Press.

Haq, G. and A. Paul (2012) *Environmentalism since 1945*. London and New York: Routledge.

Heater, D. (2004) *A Brief History of Citizenship*. Edinburgh: Edinburgh University Press.

Joppke, C. (2007) 'Transformation of Citizenship: Status, Rights, Identity'. *Citizenship Studies*, 11(1): 37–48.

Online resources

Rachel Carson, 'A Fable for Tomorrow' (from *Silent Spring*). Available at: http://core.ecu.edu/soci/juskaa/SOCI3222/carson.html

NASA Climate Reel: A collection of NASA's best videos of climate change. Available at: http://climate.nasa.gov/climate_reel

The Human Impact on This Earth (uploaded 28 March 2012). Duration 5.33 mins. Available at: www.youtube.com/watch?v=3zyizEz9XUs

The Limits to Growth (uploaded 17 February 2011). Duration 1.26 mins. Available at: www.youtube.com/watch?v=9y46hJCaHVQ

Why Citizenship Matters (uploaded 23 April 2009). Duration 6.09 mins. Available at: www.youtube.com/watch?v=8XfPdtXSLBk. Also available at: www.bbc.co.uk/learningzone/clips/why-citizenship-matters/6132.html

Student activity

Explore the intersections between environment and citizenship. This activity is designed to spark thinking and conversation about environment and citizenship, and to begin exploring the complexity and significance of those connections and intersections.

Watch the BBC video 'Why Citizenship Matters' (2009). In the story, Hamza, a student, explores why citizenship matters in his daily life. The film is divided into two parts, showing how, depending on how he lives his life, Hamza's daily routine can have positive or negative consequences for the environment and the community. Note how two of the three aspects of citizenship explored in the story explicitly have to do with environmental issues and behaviour, suggesting a strong link between being a good citizen and being a good environmental citizen. The third issue brings up the notion of global citizenship (explored later in the text).

After watching the video, and based on your current understanding of citizenship and the environment, try to map in written form what you think the fundamental key words, themes and issues that connect citizenship and the environment are. Think of questions such as these: which environmental issues have a direct impact on the traditional conception of citizenship and on any of its elements? Which aspects or elements of citizenship are framing environmental issues and concerns? Is it possible to be a good citizen in the twenty-first century without taking environmental issues into account? This activity should anticipate some of the key themes and intersections between environment and citizenship that will be explored in the remainder of this text.

Discussion questions

- What do you understand by the environment? Do you see it as synonymous with nature? Do you consider humans to be part of the environment and/or part of nature?
- How much of your environmental awareness comes from accidents and disasters? Should these be important in shaping our environmental views and values? Can they distort those views?
- What does the term citizenship mean to you? Which aspect of citizenship is more important to you: membership, rights or duties?
- Do you see yourself as a full citizen? Do you think that you have access to all the entitlements of citizenship? If so, what are these? If not, what do you think you are missing out on?

2 Introducing citizenship theories

There is little agreement about the meaning of citizenship in the rapidly changing social, economic and political conditions of late modernity.

Nick Ellison (1997)

This chapter builds on the portrait of citizenship provided in Chapter 1. The chapter cannot do justice to the diversity and complexity of citizenship theories, let alone the many variations formulated by the hundreds of authors who have theorized about citizenship over the centuries. Thus, we will pay special attention to those theories that will facilitate the fullest appreciation of the impact of environmental issues and ideas on contemporary formulations of citizenship, which is the specific purpose of Chapter 3. We will explore six theories of citizenship, organized into three sections: (1) classical theories (liberal and republican); (2) pluralist theorists (feminist and multicultural); and (3) globalist theories (cosmopolitan and neoliberal). We begin with the two theories that still dominate the field of citizenship: liberal and republican.

Classical theories

The classical theories of citizenship, republican and liberal, share a long history that goes back to ancient Greece and Rome. They deserve this label not only because they emerged during the classical period of history, but also because they have set the terms of the debate on the subject of citizenship ever since (Pocock 1992). In other words, they remain the dominant or mainstream theories when it comes to framing debates and issues on citizenship. The current content of republican and liberal citizenship has significantly changed since their ancient origins, yet their structure remains remarkably similar, and thus it is both useful and important to take a historical look at their origins and general trajectories. This will help us see which traits can be considered essential and which might be considered contextual. To that end, we will begin the exploration of each of the two models with a brief historical overview, and then provide an account of the essential traits of each theory. We begin with the model that is synonymous with the birth of citizenship: republican citizenship.

Republican citizenship

Historical overview

Republican citizenship emerged and developed in the city-states of ancient Greece, especially Athens and Sparta, and was first theorized by Aristotle (Figure 2.1), most notably in his classic text *Politics* (c.330 BCE). Aristotle regarded humans as political animals (*zoon politikon*) whose nature it is to live in political communities, and he regarded citizenship

Figure 2.1 *Greek philosopher, Aristotle (384–322 BCE)*

Source: Can Stock Photo Inc.

as essential to the realization of our humanity. Citizenship entailed first and foremost the exercise of political participation. This was conceived not as a means to an end, but as an end in itself – a necessary part of realizing one's humanity. Aristotle defined citizens as 'all who share in the civic life of ruling and being ruled in turn', that is, all those who simultaneously rule (make law) and are ruled (by the law). This definition of citizenship as self-rule (or self-government) is still used to differentiate citizens (those who rule and are ruled) from subjects (those who are ruled by others).

The foremost expression of Aristotle's conception of citizenship was ancient Athens. The essence of Athenian citizenship was dedication to public life. The emphasis was on the obligations of citizens toward the community rather than on the rights of the citizens. Athenian citizens were expected to pay taxes, obey the laws and provide military service. But, above all, they were expected to participate in the everyday running of the community, mainly by speaking and voting in the assemblies, holding political office and administering justice. This gave citizenship strong normative connotations. In the words of Pericles, those who did not concern themselves with the matters of the community were regarded not as people who mind their own business, but as people who have no business at all in the community (Manville 1990: 15).

Citizenship was a highly demanding occupation that required citizens to sacrifice their private interests and devote the better part of their time and energy to public affairs. Such a demanding model of citizenship was made possible by a sharp division of labour built upon a rigorous separation between public life (the life of the *polis*, the city-state) and private life (the life of the *oikos*, the household). Citizens were expected to dedicate their lives to the running of the community (the public sphere), leaving the running of the household (the private sphere) to women and slaves. This meant that 'citizens could only dedicate themselves to public life because their private lives were serviced by others' (Bellamy 2008: 35). In addition, most inhabitants were excluded from citizenship because they were not considered fit to rule, for a range of reasons: slaves (born to serve), women (irrational), children (immature) and *metics* (foreign residents, and therefore of questionable loyalty). In essence, citizenship was restricted to adult, free males of Athenian descent, roughly the equivalent of one-tenth of the population of Athens.

The other major Greek influence in the origins of republican citizenship was Sparta. This city-state, famous for its fierce rivalry with Athens, was also renowned and admired for the selfless devotion of its citizen-soldiers, who symbolized the ultimate realization of the republican ideal in ancient Greece. Spartan citizenship has been described as 'an intensification of the Athenian notion of public service' (Riesenberg 1992: 8). But, unlike democratic Athens, Sparta was an authoritarian regime, and thus a reminder that citizenship, whilst typically associated with democracy, has been historically compatible with different forms of government, including autocracies, monarchies and empires. Spartan citizenship was also highly exclusive, in this case to free, adult males of Spartan descent, and full citizenship was only earned after completion of military service, when aged 30.

The Greek model of citizenship inspired the civic republicanism of ancient Rome. In fact, the term republic comes from the Latin *res publica* (the public affairs, or commonwealth), the expression that gave name to the Roman Republic. This period produced the other major classical author on republicanism, Cicero. His work emphasized the importance of virtue for the exercise of good citizenship, most notably in his classic text *On Duties* (44 BCE). Citizenship lost its association with ethnicity but, on the whole, the fundamental structure of republican citizenship remained the same, retaining its fundamental association with civic duty and political participation.

The republican conception of citizenship was largely abandoned following the decline of the Roman Republic and was not fully revived until the period of the Renaissance, in the city-states of Venice and Florence. The revival of the republican ideal was articulated by Niccolò Machiavelli, most famously in *Discourses* (1517) and *The Prince* (1513). Machiavelli presented self-rule as the key to citizenship and advocated the need to cultivate civic spirit. He highlighted the importance of military service, which he considered necessary to defend the republic against internal discord and external attack. Indeed, his ideal citizen was the citizen-soldier. Machiavelli emphasized the moral sense of citizenship (the *virtú*, or virtue required to be a good citizen), but also its militaristic dimension, revealing his admiration for Sparta.

The republican tradition was modernized during the period of the Enlightenment, most notably by Jean-Jacques Rousseau. His work, particularly *The Social Contract* (1762), inspired the French Revolution, and underpinned the foremost modern document of republican citizenship, the Declaration of the Rights of Man and the Citizen (1789). The social contract is and has remained central to republican and liberal citizenship. Indeed, since the revolutionary era both models have coexisted under the modern conception of national citizenship. However, liberal citizenship has become increasingly dominant over time, especially during the twentieth century. The loss of civic duty and community spirit, and the decline in political participation and public service that has resulted from the ascendancy of liberal citizenship have been lamented by many authors, amongst them Hannah Arendt (1958), Robert Putnam (2000) and Michael Sandel (2012).

Essential traits (and critiques)

The Collective vs. *The Individual.* Republican citizenship privileges the collective over the individual and emphasizes the need to create a common bond between citizens. The citizenry must form a community, not be merely a collection of individuals. The primacy of the collective makes patriotic solidarity and loyalty essential to republican citizenship. Their cultivation creates a strong collective bond – the sense of unity and integrity that makes living together possible (Hanasz 2006: 285). In the republican model, the individuals who make the social compact are born anew as citizens whose identities are indissolubly bound up with the community of which they are a part. The foremost responsibility of each citizen is to sustain the community itself. Republican citizenship stresses the importance of social capital, that is, the cultivation of relationships, connections and social networks amongst individuals, from which trust and reciprocity arise. In the republican model, individual freedom, security and happiness derive from belonging to a political community. This has always been synonymous with a bounded political community (e.g. the city-state, the nation-state). The exclusive character of republican citizenship when it comes to outsiders has also operated internally, with particular identities determining the status of different groups at different times in history. In modern times, republican citizenship has focused on the need to create a political community that, whilst still bounded (and exclusive to insiders) bridges differences of class, gender and culture between its members. In any case, the primacy of the collective remains intact, and the political community continues to be predominantly associated with the nation-state.

The Public Good vs. *Private Interests.* Republican citizenship is based on the distinction between the public and the private sphere but, in contraposition to liberal citizenship, prioritizes the public good at the expense of private interests. The ideal citizen is the one who participates in public life with the purpose of serving the common good. This model of citizenship is built on the notion of civic virtue, which demands that, as citizens and members of the public, 'people must be prepared to overcome their personal inclinations and set aside their private interests when necessary to do what is best for the public as a whole' (Dagger 2002: 147). The development of civic virtue requires public education that can instil standards of public morality and the notion of the common good. Obeying the law is necessary, but not sufficient. The good citizen is the public-minded, spirited, virtuous citizen, not just the law-abiding citizen. In the republican tradition, citizenship requires an ethical commitment to the public good.

Political Participation *vs. Legal Protection.* Republican citizenship is created through the participation in public affairs, understood first and foremost as political participation. This translates into a conception of citizenship as action, where collective action provides for individual freedom: 'to be free and to act are the same' (Hanasz 2006: 295). The citizen is not only a political being, but also a political agent. The good republican citizen cannot be a political spectator. In the early days of republican citizenship, political participation took the form of direct democracy, but in modern societies it takes place mainly through representative democracy. However, republican citizenship demands that we engage in politics not just by exercising our right to vote but through participation in political parties, local assemblies, and other actions that come under the banner of deliberative democracy (e.g. public debates and discussions). The emphasis on political participation usually leads to the association of republican citizenship with active citizenship. The good republican citizen is the politically active citizen.

Duties and Obligations *vs. Rights and Entitlements.* Republican citizenship emphasizes duties and obligations over rights and entitlements. Indeed, republican citizenship is often referred to as duties-based citizenship. The good republican citizen is the one who recognizes that citizenship is a matter of duties, obligations and responsibilities towards the community, and discharges those responsibilities accordingly. These include day-to-day obedience to the law, paying taxes, military service or other form of public service, and playing a well-informed and public-spirited part in the affairs of the community. Citizens earn their status by attending to their public duties and responsibilities. To assist with this, the republican tradition has always been concerned with fostering the public virtues that enable and inspire citizens to fulfil those duties. The good republican citizen is the dutiful citizen – those who contribute to the overall good of the community by fulfilling their obligations towards their fellow citizens.

Figure 2.2 *'Your Country Needs You'*

Source: Cover courtesy of *Prospect Magazine*.

Critiques of republican citizenship. The main critique levelled at republican citizenship relates to its exclusive tendencies and its demanding nature. Whilst republican citizenship has made progress in terms of internal inclusion, its continued association with bounded political communities (currently the nation-state) remains a point of criticism, mainly from globalist theories. The second major critique relates to the potentially coercive nature of republican citizenship that comes with its demanding nature. Historically, this has translated, for example, into compulsory military service (still a reality in many countries). The coercive side of the republican tradition is also illustrated in compulsory civic service, the most common being jury duty, but which can include a wide range of community service schemes (e.g. visiting the elderly, helping out in schools, mentoring young children and renovating public spaces). These schemes are sometimes proposed to strengthen citizenship in liberal democracies, as is currently the case in Britain and Australia (Figure 2.2). The two critiques come together in the area of education for citizenship, which, critics argue, transforms education into a form of indoctrination that promotes a patriotic community and produces docile citizens. The exclusive, demanding and potentially coercive character of republican citizenship can limit the scope of individual citizens to pursue their own interests. This critique of republican citizenship comes, first and foremost, from those who embrace the other classical and mainstream citizenship theory: liberal citizenship.

Liberal citizenship

Historical overview

The origins of liberal citizenship can be traced back to ancient Rome. As a result, liberal citizenship is also referred to as the Roman model. Roman citizenship resulted from the struggle of the plebeians (or lower classes) to obtain rights against the patricians (or upper classes) who controlled the central institution of political power during the Roman Republic, the Senate (Bellamy 2008). The ongoing class struggle gave politics and citizenship a much more instrumental character than the ideal of public-spirited service associated with the Greek model. In addition, citizenship became more closely associated with legal protection and jurisprudence (the rule of law) than political participation and self-government (the making of laws). Whereas in ancient Greece the citizen was primarily a political being (i.e. he who shared in the running of the community), in ancient Rome the citizen was mainly a legal being (i.e. he who was free to act by law and to expect the protection of the law). The essence of this new model was captured in the three great precepts of Roman law for its citizens: 'to live honourably, not to cause harm, to give each their due'. These precepts, derived from the *Corpus Iuris Civilis* (Body of Civil Law), also known as Justinian's Code (529–534), the most influential summary of Roman legal doctrine, could almost be taken as the script for early modern liberal citizenship (Burchell 2002: 96).

The development of Roman imperial rule eroded the active and political character of citizenship that could still be found during the Roman Republic, which was influenced by the Greek model. With political power concentrated in an oligarchy based in the capital of the empire, citizenship no longer required political participation. The conquered peoples were granted a reduced version of Roman citizenship, namely citizenship without the vote (*civitas sine suffragio*), that is, legal but not political rights. Roman imperial citizenship did not imply active citizens who exercised political power, but passive citizens under the legal protection of the Roman Empire. Rights were granted to citizens, not legislated by them. The legal conception of citizenship facilitated its gradual extension to new peoples, as the power of Rome expanded, first across the Italian peninsula and then across Europe and into Asia and Africa. Rome demanded political allegiance, obedience to the law and payment of taxes, and, in return, citizens across the empire were free to carry out their private lives under the legal protection of the Roman Empire. Even the most common of civic obligations (military service) dwindled and finally disappeared over the imperial period, as armies were raised first on a regional and then on a purely professional basis. Roman citizenship became essentially a legal status that provided rights to legal protection by Roman soldiers and judges, in return for allegiance to Rome. The new model revealed the law and citizenship as a means for individual protection, but also as a tool of imperial rule. Conquered peoples had something to lose if they decided to rebel against the empire, not least because treason carried the death penalty (Adams and Porter 2008).

In its modern sense, liberal citizenship has its origins in the period of the Enlightenment, particularly in the liberal political theory developed by John Locke in *Two Treatises of Civil Government* (1690). Locke's liberalism is built upon his theory of natural rights – the notion that 'every man' (and at the time it was literally every 'man') has the 'free and equal right' to preserve 'his life, liberty, and estate'. Locke emphasized the rights against the state, but also a specific set of rights that has become central to liberal citizenship: property rights. He stated: 'government has no other end but the preservation of property'. His work reveals the close link between liberalism and capitalism and illustrates the notion that liberal citizenship is 'a political expression of capitalism' (Heater 1999: 7). The central place of property in the liberal conception of citizenship was never clearer than when the right to vote and the right to hold office were exclusive to men of property. The franchise was gradually uncoupled from wealth and property, and extended to other groups, including women and the working class. However, the continued association between liberalism and capitalism produced a tension between the increasing political equality and the persistent socioeconomic inequality. The most influential liberal theorist who attempted to address this tension was T.H. Marshall (Figure 2.3), most notably in his essay 'Citizenship and Social Class' (1950).

Figure 2.3 *T.H. Marshall, author of 'Citizenship and Social Class' (1950)*

Source: Wikimedia Commons.
Author: Library of the LSE.

Marshall noted how many members of the working class had become enfranchised but poverty and insecurity prevented their full participation (as citizens) in the community. His answer was to add a third layer of rights: social rights. These rights were designed to soften the negative impact of the economic inequalities created by capitalism and the class society. This came to be known as social liberalism – to differentiate it from classical liberalism, but not to be confused with socialism, which emphasizes duties and the collective, and is therefore a manifestation of republican citizenship, observable today in China and Cuba. Social liberalism emphasizes the need for social rights as a precondition for political equality and democracy. Social rights became the cornerstone of the 'welfare state', a social and political model that witnessed its heyday in Western Europe, between 1945 and the late 1970s. The social turn was criticized from within the liberal tradition by eminent figures such as Friedrich von Hayek and Robert Nozick, who argued that social rights demanded a much more substantial role for the state and would strip away individual autonomy and freedom, thus militating against the essence of liberalism. This thinking underpinned one of the most influential texts, Hayek's *The Road to Serfdom* (1944). The call for a return to classical liberalism and the notion of a minimal state was embraced by the conservative governments of Ronald Reagan in the United States and Margaret Thatcher in the United Kingdom in the 1980s. Their market-oriented policies underpinned the transition from the welfare state to the 'workfare state' and signalled the emergence of the citizen as consumer that would come to be associated with neoliberal citizenship.

Essential traits (and critiques)

The Individual *vs.* ***The Collective.*** Liberal citizenship is defined first and foremost by the primacy of the individual over the collective. Humans are conceived as atomistic, rational agents whose existence and interests are ontologically prior to the community. The purpose of citizenship is to ensure the freedom of all its members to pursue their interests and realize their capabilities as individuals. In the liberal tradition, the individual is not transformed or born anew by citizenship, but remains, first and foremost, and always, an individual. The centrality of the individual translates into a thin sense of collective identity. What brings individuals together as citizens and binds the political community is not any sense of collective identity, community sentiment or collective solidarity, but the rule of law. Identity is something personal, irrelevant to public life. Thus, everybody, irrespective of their identity, should have an equal place in the political community. This makes liberal citizenship seemingly neutral and inclusive regarding matters of identity. However, that inclusion is still limited to those who formally belong to the bounded political community of the nation-state – although there is a sense in which liberalism can easily embrace the cosmopolitan ideal of global citizenship. In any case, the individual remains the core subject of liberal citizenship.

Private Interests vs. *The Public Good.* Liberal citizenship is also based on the distinction between public and private, and retains the association of citizenship with the public sphere. The two spheres are kept distinct, but in direct contrast to the republican tradition liberalism gives primacy to the private sphere. The realm of the citizen is the public sphere, but this is a limited space in which individuals can participate and contribute as citizens if they wish, but are under no obligation to do so. The state exists to protect individual rights and facilitate the pursuit of one's private life and interests, to enable the pursuit of individual happiness. The state has no role to play in deciding over private matters such as notions of the good. This often translates into support for private rather than public education. In the liberal tradition, education in general and education for citizenship in particular should be neutral, avoiding the promotion of particular identities and notions of the common good. In other words, liberals prefer an education in which each individual can choose their personal identity and construct their own vision of the good life. The only constraint is the rule of law and the general principle of not harming others in the pursuit of one's interests, so that others can also pursue their own personal interests.

Legal Protection vs. *Political Participation.* The essence of liberal citizenship is legal protection. This translates into a conception of citizenship as legal status. The protection offered by citizenship – through the state and from the state – opens up the possibility for individuals to act as citizens in the public sphere, which is the space where citizenship is exercised. In the liberal tradition, however, the individual has no obligation to participate. Participation, political or otherwise, is a right, not a duty. The primary purpose of the legal protection of individual rights is to enhance the private sphere so that individuals can enjoy and pursue their personal interests in a more secure and comfortable environment. In addition, liberal citizenship means protection from the authorities themselves. Classical liberalism emphasizes the freedom of the individual citizens from the constraints of the state. The ideal citizen is the autonomous individual, not the active citizen. Individuals can be as active or passive as they wish. The citizen is, above all, a legal being, not a political being. The good liberal citizen is the law-abiding citizen.

Rights and Entitlements vs. *Duties and Obligations.* Liberal citizenship is strongly associated with rights, and is often referred to as rights-based citizenship. The liberal citizen is, first and foremost, a bearer of rights. There are, however, important debates amongst liberals about which rights to include as well as about their relative value. The relative importance attached to rights against the state (or negative rights, such as civil rights) and rights provided by the state (or positive rights, such as social rights) remains central to internal disputes amongst classical and social liberals. The centrality of rights does not mean that liberal citizenship is mute on responsibilities. These include: obeying the law (provided the law is just), paying taxes (in accordance with the law), and monitoring the government (to ensure it complies with the rule of law). The latter duty refers to a liberal conception of 'citizenship as vigilance' that shares some affinity with republican citizenship (Enslin and White 2003: 121). However, liberal duties (other than law abidance) are framed in terms of voluntarism rather than civic obligation. The focus of liberal citizenship remains squarely the protection of individual rights.

Critiques of liberal citizenship. Liberal citizenship is criticized for enabling and fomenting individualism and social inequality by not paying sufficient attention to fostering public virtues and the common good. This, critics argue, produces an impoverished version of citizenship in which citizens are reduced to atomized, passive bearers of rights whose freedom consists in being able to pursue their selfish interests (e.g. Lister 1997a). The emphasis on individual rights and the voluntary approach to public life often lead to labelling liberal citizenship as passive citizenship. This label has some merit, but we must keep in mind that liberal citizenship is activated when rights are challenged or violated. In other words, the liberal tradition demands that liberal citizens become active when the state fails to fulfil its side of the social contract – to guarantee individual rights. The active aspect of liberal citizenship is also visible when we look at the struggle for different rights. The gaining of rights has always been through political actions. The other major critique of liberal citizenship relates to the inequality generated by its close association with capitalism, the centrality of property rights, and the selfish pursuit of riches. The criticisms noted thus far tend to come from the republican tradition (including its socialist variation). Liberal citizenship is also criticized for its universalist assumptions regarding the identity of the citizen and for its modern association with the nation-state. These criticisms, levelled also against modern republican citizenship, come from alternative theories, to which we now turn. We begin with those that challenge classical theories from below: pluralist theories.

Pluralist theories: difference matters

Pluralist theories challenge the universalism conventionally associated with the classical theories of citizenship. Pluralists place power and identity/difference at the centre of their analysis. They show how certain groups have been and still are excluded from full citizenship because they are deemed to be 'different' (e.g. women, sexual minorities, indigenous peoples, ethnic minorities and people with disabilities). The classical theories, they argue, have operated with a specific and particular conception of the citizen: male, white, heterosexual, affluent, able-bodied and adult. Their supposed universalism is in fact the particularism of the dominant group presented as universal. In other words, pluralists refute the notion that citizenship offers equal status to all citizens by pointing out how some citizens/groups are more equal than others due to their identity (e.g. men are more equal than women). Pluralist critiques come in two main forms: feminist critiques, focused primarily on the issue of gender, and the discrimination suffered by women; and culturalist critiques, focused on the issue of culture, and the discrimination suffered by ethnic minorities and indigenous groups. We begin with the theory that has posed the most important challenge to the classical theories: feminist citizenship.

Feminist citizenship: gender matters

Feminist citizenship is underpinned by feminism, a rich and diverse theoretical and ideological field that cannot be reduced to a single model (e.g. liberal, socialist, radical). However, at the heart of all feminisms is the notion that gender matters, not least because it structures much of social and political relations, including citizenship. In general terms, feminists tend to promote a rights discourse that aims to place women on an equal footing with men. This discourse can be traced back to the first and most venerated work in defence of women's rights, Mary

Wollstonecraft's *Vindication of the Rights of Women* (1792). Since then, feminists have focused most of their analyses and their actions on obtaining equal rights with men (Figure 2.4), but over time they have also argued for additional rights based on sexual differences (e.g. reproductive rights, right to abortion, maternity leave). The feminist struggles for full citizenship have translated into a rich vein of academic work that has documented and theorized the historical exclusion of women from citizenship, and formulated alternative models to promote their full citizenship.

Feminist theorists have documented women's exclusion from citizenship both in terms of membership and in terms of rights and duties. On the one hand, they have shown how women have been historically excluded from formal citizenship or how their inclusion was dependent on the citizenship status of their fathers or husbands (e.g. Nicolòsi 2011). On the other hand, they have shown how women were either subjects (ruled, but not ruling) or second-class citizens (with limited and/or derivative rights), and often could only exercise those rights through male representatives. In short, theirs was a dependent status and/or a derivative citizenship. Their protection was limited, and their ability to participate minimal or

Figure 2.4 *'The Suffragette that Knew Jiu-Jitsu. The Arrest'*

Source: Wikimedia Commons. Cartoonist: Arthur Wallis Mills.

non-existent. This has changed significantly in recent times and in most places, but not completely, and not everywhere. There are still cultural and structural obstacles that continue to prevent women's full access to citizenship in many (if not most) countries. In addition, feminist theorists have begun exploring intersections between gender, age, class, sexuality, ethnicity, nationality, racialization, dis/abilities and other social divisions that provide a more complex portrait of the gendered nature of citizenship (Oleksy *et al.* 2011). Their studies reveal that citizenship continues to be a gendered institution, both in the Global North and the Global South (Lister 2008).

A theory of power and the politics of time

The main contribution feminism has made to citizenship derives from its theory of power. Feminist theorists have shown how power operates at all levels of society, not just in the public domain (the conventional sphere of politics, and traditionally associated with citizenship), but also in the private domain (the traditional sphere of personal relations, and commonly excluded from citizenship). It is this insight that inspired the famous feminist slogan: the personal is political. This theory of power challenges the private/public divide that has been central to the institution and practice of citizenship. Feminist theorists have exposed this rigid divide as mythical, noting that the two spheres 'are, and always have been, inextricably connected' (Okin 1998: 69). First and foremost, these theorists have revealed how participation in public life can only function as it does if significant domestic service is provided. In addition, they have shown how public policies shape the private conditions of life (e.g. domestic violence, child rearing). Last but not least, feminist theorists have exposed the gendered nature of the public/private divide that prevents women from gaining full access to the realm of citizenship (e.g. Lister 1997a).

Significantly, feminist theorists argue that the gendered public/private divide is not the unfortunate consequence of a particular conception of citizenship, but constitutes the very essence of the mainstream notion of citizenship. The exclusion of women from formal citizenship throughout both modern and ancient history – they argue – has not been incidental or accidental, but 'a necessary condition for citizenship' (Lister 1997b: 68). Citizenship has only been possible because the gendered division of labour has kept women in charge of the domestic sphere, allowing men the time to exercise the life of the citizen. This brings up a significant feminist insight: 'time is a necessary resource for the practice of citizenship' (Lister 1997a: 201). The amount of time women spend, for example, working as carers and performing domestic tasks limits their ability to participate in public life, the dimension of life traditionally associated with citizenship.

The way forward: feminist dilemmas

There is a debate amongst feminists about the value of citizenship as a concept, given its long historical association with sexism and patriarchy. Some have argued for abandoning the concept and trying to formulate conceptions of politics which are not constrained by the historical baggage of citizenship. Others have embraced citizenship and defend the concept as a potentially radical ideal that can inform both theoretical analyses and practices of membership in political communities. This second group sees citizenship as a promising strategy for challenging gender inequality. The problem is not citizenship but the historical context in which citizenship operates: 'context is all' (Dietz 1987). In a different (non-patriarchal) context, the ideal of universal citizenship contains the possibility of an all-inclusive and egalitarian articulation of citizenship. In other words, the notion of universal citizenship comes with the risk of masking gender and other particularities, but 'it also possesses the strength of its own idealism' (Riley 1992: 187).

Much of the feminist debate about citizenship centres on what to do with the private/public distinction. Some theorists have argued for abandoning the distinction altogether, noting that, since all activities have a private and a public dimension, the distinction is ultimately untenable (Yuval-Davis 1997). However, most feminist theorists advocate retaining the distinction, but reformulated. The main point of contention is whether to emphasize the private or the public dimension. Those who privilege the private dimension tend to argue that women's role as mothers and carers in the private sphere positions them in a special place to enhance the quality of public life (e.g. Elshtain 1981; Ruddick 1989). These 'maternalist feminists' argue that citizenship would benefit from an injection of the values they associate with women (e.g. caring and nurturing). In other words, the values of the home (the private sphere) would improve the politics in the agora (the public sphere). Theorists who privilege the public dimension tend to argue for redrawing the boundaries between the private and public spheres in ways that will facilitate women's participation in public life. This would require, for example, treating certain activities that take place in the private sphere (e.g. caring) as a matter of public service, and/or democratizing the private sphere so that men and women can have the time to contribute equally to public life. The reform of the public/private divide along these lines calls for a double democratization, of public life and family life (Phillips 1993).

The other major debate relates to the equality/difference dilemma; between visions based on gender equality (i.e. liberal-minded) and those based on sexual difference (i.e. pluralist-minded). Feminist citizenship can mean calls for equal treatment (e.g. equal pay), permanent differential treatment (e.g. reproductive rights) and/or compensatory differential treatment (e.g. affirmative action). Liberal feminists tend to argue that special treatment would prevent women from competing on equal terms in the labour market, ultimately forcing them into economic dependency on men, even if unwittingly. Pluralist feminists tend to argue that gender equality requires addressing issues that are specific to women (e.g. pregnancy). Anne Phillips tries to circumvent this dilemma by arguing that gender-neutral is the ideal, but until genuine gender equality is achieved, particularly regarding the sexual division of labour in the private sphere, gender-differentiated strategies are necessary in order to redress the imbalance (Phillips 1991). Ruth Lister attempts to reconcile the universalism that lies at the heart of citizenship with the demands of a politics of difference by grounding citizenship on the notion of 'differentiated universalism' (Lister 1998). This formula aims to realize the inclusive potential and the egalitarian promise of citizenship, on the basis that such a promise can only be delivered through a politics of difference.

Critiques of feminist citizenship

Feminist accounts of citizenship can be criticized for paying excessive attention to gender. This is, of course, a critique that can be applied to any theory that pays special attention to a particular trait, be that gender, class or ethnicity. The main specific criticisms of feminist citizenship are associated with one term: essentialism. This critique takes two main forms: that feminist theorists essentialize people (particularly women) by making gender essential to their identity; and that they operate with an essentialist notion of gender that associates women with certain traits (e.g. motherhood and caring) and men with others (e.g. aggression and fighting). The feminist response has been to rethink gender as a social construction, with no fixed meanings, even as 'performance' (Butler 1994). In other words, feminists now operate with conceptions of identity/difference that embrace the fluidity, fragmentation, contingency and instability of people's actual identities. Feminists have also created formulas that embrace conceptions of politics that are intersectional and transversal (i.e. bringing together gender, culture, ethnicity and class) (e.g. Yuval-Davies 2011). These formulas, however, come with problems of their own, not least their practicality, especially when it comes to making the case for differentiated citizenship. If gender is a construct or a performance, on what grounds can we establish criteria to assert gender-specific rights and formulate gender-specific policies? And how do we create effective political coalitions that bring together such a disparate array of groups and individuals? Here, citizenship stops being a mere legal or political status to become an infinitely reconstructible 'articulating principle' (Ellison 1997: 697). But how are we to operationalize such a fluid conception of citizenship? Similar criticisms, albeit regarding culture rather than gender, are levelled against the other pluralist theory: multi/cultural citizenship.

Multi/cultural citizenship: culture matters

The other significant departure from thinking about citizenship in a universalistic way comes from the emergence of cultural citizenship. This articulation of citizenship is built on the notion that culture matters to people, that it is essential to human life, including to the life of the citizen. Culturalist theorists challenge the liberal notion that the public sphere should be value-neutral and that cultural identity should be relevant only in the private sphere. Instead, culturalist theorists argue that culture always has been (and should be) essential to public life. They note that cultural identity/difference has played a significant role in the history of citizenship, albeit typically to the detriment of cultural minorities, from the *metics* of ancient Greece to the indigenous peoples of modern Australia. In addition, they argue that, in the context of multicultural societies – and to some extent all modern societies are multicultural, i.e. composed of a diversity of cultures – it is impossible to be neutral regarding cultural matters. The neutrality of the state enables the dominance of the majority cultures. This is the context in which cultural citizenship enters the scenario, striving to promote the recognition and empowerment of cultural minorities. The most innovative and influential formulation of cultural citizenship developed to address these issues – and the main one to be explored here – is multicultural citizenship.

A theory of minority rights

Multicultural citizenship is a theory of minority rights formulated in the context of liberal democracies. The most influential articulation of multicultural citizenship is that of Canadian political scientist Will Kymlicka, best formulated in *Multicultural Citizenship* (1995). Kymlicka argues that multicultural states should have both universal rights, assigned to individuals regardless of group membership, and certain group-differentiated rights or 'special status' for cultural minorities. He points out that many government decisions, including those on languages, internal boundaries, public holidays and state symbols inevitably involve 'recognizing, accommodating, and supporting the needs and identities of particular ethnic and national groups' (Kymlicka 1995: 108). This effectively means that, in multicultural societies, the liberal policy of 'benign neglect', far from being neutral, favours majority cultures. In these societies, justice demands that traditional human rights be supplemented with minority group rights. The specific rights would vary depending on each specific case and the type of cultural group (e.g. immigrants, national minorities and indigenous peoples), and they could include, amongst others, the following: affirmative action, language rights, religious exemptions, media representation, regional autonomy, land claims, customary law and self-government rights (Kymlicka 2010: 260).

The emphasis on rights situates multicultural citizenship in the sphere of the liberal tradition, albeit complemented with a communitarian sentiment derived from the importance attributed to the cultural community. Multiculturalist authors of a liberal disposition, such as Kymlicka and Chandran Kukathas, have made a point of balancing the individual and the collective, the personal and the cultural, by articulating mechanisms that protect individuals from potential oppression by the group – in the form of the 'right of dissent' (Kymlicka 1995) and the 'right to exit' (Kukathas 2007). The first implies rejecting internal restrictions that limit the rights of group members to question traditional authorities and practices. The second implies the right to leave the group. The extent to which minority group rights help to facilitate the social integration and political engagement of cultural minorities or are a recipe for social isolation, ghettoization and political fragmentation remains a matter of contention. The debate has intensified in recent times, particularly following the 9/11 attacks and the subsequent backlash against Islam, and the populist reactions against migration that followed the Global Financial Crisis (GFC).

Indigenous citizenship

Multicultural citizenship contemplates the rights of indigenous peoples. However, for the most part, the concept has been used to articulate policies designed to integrate migrants and ethnic minorities, mainly in the countries of the Global North. In this context, it is hardly surprising that cultural citizenship has taken a different emphasis in the Global South. Here, its focus has been on the struggle over indigenous rights. This has translated, for example, into the formal inclusion of indigenous people as citizens in countries where they had been historically excluded from citizenship, or included as dependent citizens, as was the case in Brazil prior to 1988 (Ramos 2003). The lack of sufficient attention to indigenous rights in official multicultural frameworks has also produced calls for indigenous citizenship in countries of the Global North where indigenous populations continue to suffer from widespread poverty and discrimination, as is the case in Australia (Rowse 2000). In addition, indigenous citizenship has gained international recognition, particularly through the United Nations Declaration of the Rights of Indigenous Peoples (UNDRIP). This text reflects the mainstream conception of indigenous citizenship – and one that is of particular relevance here, given the close association it establishes between indigenous peoples and indigenous rights and the natural world.

The starting point of the declaration is the marginalization of indigenous peoples, associated with the legacy of colonialism: disease, poverty, violence, starvation, slavery and forced sterilization. The text acknowledges the diversity of indigenous peoples, but notes that for all their diversity, as communities living in intimate connection with the natural world, indigenous peoples share similar worldviews, which include: the intrinsic value of nature, the sense of nature as teacher, the interdependence of all natural life and the responsibility towards nature and all life (including often a spiritual connection with the land). The declaration then contemplates a long list of rights, including amongst many others: the right to self-determination; the right to their distinct political, legal, economic, social and cultural institutions; the right to their traditional medicines and to maintain their health practices, including the conservation of their vital medicinal plants, animals and minerals; the right to the lands, territories and resources which they have traditionally owned, occupied or otherwise used or acquired; the right to the conservation and protection of the environment and the productive capacity of their lands or territories and resources. The explicit and strong link made between indigenous peoples and the natural world and some of the rights contemplated to protect that link refer to the notion of multicultural environmental citizenship that will feature later in the text.

Critiques of (multi)cultural citizenship

Culturalist theories of citizenship have been criticized for overstating the significance of culture. Similarly to feminist citizenship, culturalist citizenship has often been accused of essentialism. This critique takes two main forms: that culturalist theorists essentialize people (particularly cultural minorities) by making culture essential to their identity; and that they tend to operate with fixed and static notions of culture. Thus, for example, multicultural citizenship has been criticized for treating cultural minorities as puppets or captives of their cultures and cultural identities. In addition, feminist theorists have argued that, given the patriarchal nature of most (if not all) cultures across the world, 'multiculturalism can be bad for women' (Okin 1992). In response, some theorists have tried to address these critiques from a multicultural and feminist perspective. The most interesting attempt at this is the work of Anne Phillips, who argues that we need to extend to culture the same qualifications of fluidity and relativity that have become the norm regarding gender (and class). This would produce a regime of strong toleration that dispenses with static and strong notions of culture: 'multiculturalism without culture' (Phillips 2007). In fact, increasingly, the theoretical and empirical work of multicultural scholars indicates that identity is multiple, changing, overlapping and contextual, rather than fixed and static. However, once again, these ideas come with their own problems, especially when combined with the fluidity of groups, and the fragmentation and individualization of cultural identities. These formulas can dissolve identity and groups to the point that it becomes impossible to operationalize cultural policies. How can states grant rights to 'groups' with boundaries so porous and identities so fluid? Perhaps the best way forward could be replacing cultural differences with common global identities. The articulation of these approaches can be found in the challenge to traditional citizenship from above, that is, from globalist theories.

Globalist theories: beyond states

The other major challenge to mainstream theories of citizenship comes from globalist theories, that is, from accounts of citizenship critical of the nation-state centrism of modern citizenship. The challenge originates mainly from the acceleration of globalization in the second half of the twentieth century. The changes to social, economic and political dynamics produced by globalization pose serious questions regarding the power of the nation-state. Globalization challenges the bounded nature of citizenship and poses the question of whether the nation is any longer the relevant entity for grounding rights and responsibilities. In addition, certain aspects of globalization, especially the neoliberal logic of the global marketplace, challenge the political nature of citizenship. These challenges to modern citizenship come from two main globalist theories of citizenship: cosmopolitan citizenship and neoliberal citizenship. The two are closely interrelated, at least at present, but we begin with the one that has a longer historical pedigree: cosmopolitan citizenship.

Cosmopolitan citizenship: global matters

Cosmopolitan citizenship is based on the notion that all humans belong to a single and universal community. In political terms, this community is known as the *cosmopolis*. The cosmopolitan argument is that humanity as a whole should define the primary horizon of our moral imagination and political engagement. The cosmopolitan ideal has a long history that can be traced back to ancient Greece. In fact, the word cosmopolitan derives from the Greek *kosmopolitês*, meaning 'citizen of the world', and the first documented manifestation of the cosmopolitan sentiment, that 'we should regard all men as our fellow citizens', is attributed to the Greek philosopher, Zeno of Citium. The cosmopolitan ideal was also found in ancient Rome, and has been present in religions with a universalist projection, such as Christianity, Hinduism, Confucianism and Islam (Heater 1999). In modern times, its most influential proponent was the eighteenth-century German philosopher, Immanuel Kant, who argued that only the creation of a 'universal citizenship' could prevent war. This is the argument at the heart of his famous essay 'Perpetual Peace: A Philosophical Sketch' (1795).

Cosmopolitanism survived all throughout history, but was largely confined to the world of ideas until the twentieth century. The cosmopolitan ideal achieved particular salience after the two world wars, the causes of which were at least partly attributable to the division of peoples into nation-states and the nationalist ideologies that had been used to create loyalty amongst the citizens, often by whipping up differences between nations. The cosmopolitan ideal informed the creation of the United Nations in 1945. The United Nations continues to operate as a collective of nation-states, but has produced the most explicit and influential expression of cosmopolitan citizenship to date: the Universal Declaration of Human Rights (1948). This conception of citizenship, most commonly known as global citizenship, has gained significant momentum ever since, particularly following the end of the Cold War and the arrival of the internet. The cosmopolitan impetus has been driven by the fact that globalization is compromising the integrity and power of nation-states, and as a consequence the national is seen by many as outdated or at least insufficient to account for the meaning and articulation of citizenship. 'The interdependence of the world' – it is often argued – 'implies the need for a global citizenship' (Jelin 2000: 58).

Towards a cosmopolis of Earthians

Cosmopolitan citizenship poses a radical challenge to the traditional notion of citizenship as membership of a bounded and exclusive political community. Cosmopolitan theorists see no good reason to privilege the nation-state (or any other form of bounded community for that matter) as the political space of citizenship. The nation-state, they argue, is a morally arbitrary community, since membership is determined for the most part by the morally arbitrary facts of birthplace or parentage, the so-called 'lottery of birth' (Shachar 2009). Therefore, the boundaries of the nation-state should not set limits to our moral responsibility and political commitment. Instead, we should all consider ourselves and operate as equal members of the one political community, the cosmopolis, or planet Earth.

In the present context, with the nation-state still hegemonic and political borders still very much in place, cosmopolitan theorists acknowledge that their articulation of citizenship 'implies a utopian confidence in the human capacity to exceed realistic horizons' (Falk 1994: 140). In other words, the cosmopolitan ideal requires a politics of imagination that is not bounded by the politics of the possible. That is to say, it requires 'dedicated action that is motivated by what is desirable, and not discouraged by calculations of what seems likely' (Falk 1994: 132). In this sense, whilst traditional citizenship operates spatially, grounded on the physical limits of territorial borders, 'global citizenship operates temporally, reaching out to a future-to-be-created', making those in pursuit of such vision 'citizen pilgrims', people 'on a journey to a "country" to be established in the future' (Falk 1994: 139). In short, cosmopolitan citizenship is a future-oriented project that aims to make every single one of us equal citizens of planet Earth, that is, equal Earthians.

Human rights and global duties

Cosmopolitan theories of citizenship also have an impact in significant ways on the content of citizenship. First and foremost, cosmopolitan citizenship transforms citizen rights into human rights. This implies a single, universal and equal set of rights for all human beings, irrespective of their place of birth or residence. In addition, at least presently, cosmopolitan citizenship furthers the primacy of rights over duties – a fact that reflects the strong philosophical connection between liberalism and cosmopolitanism. Indeed, from its origins in ancient times, the cosmopolitan spirit has been 'the expression of a globally oriented liberalism' (Delanty 2000: 54). To put it differently, cosmopolitanism is (at least in its prevalent form) liberalism writ large. The liberal dominance in the current articulation of cosmopolitan citizenship has been reinforced by the preeminence of civil and political rights (the classic liberal rights) over social rights in the global implementation of the human rights agenda, which has been dominated by the liberal democracies of the Global North (Burke 2010).

However, there is no reason why a republican duties-based articulation of cosmopolitan citizenship cannot emerge. In fact, some of the most influential cosmopolitan theorists have argued for increased participation and democratization of global politics (e.g. Archibugi 2008) and have produced more duties-based and justice-oriented articulations of cosmopolitan citizenship (e.g. Pogge 2002). There is also evidence of an emerging global republican sentiment in the actions of groups that emphasize global duties and participation in the political life of the global community – a sentiment often driven by environmental concerns. This approach is best illustrated by the work of non-governmental organizations (NGOs), some of which have explicitly embraced the terminology of global citizenship. Oxfam, for example, operates with the notion that we are 'citizens of the globe' with responsibilities to each other and to the Earth. Their emphasis is not on difference, but on *making a difference*. The extent to which this approach to global citizenship will lead to the emergence of a 'republican cosmopolitanism' is still very much a matter of conjecture (Hanasz 2006). However, what is clear is that the same technologies that are enhancing the prospects of a global community are also facilitating the participation of self-proclaimed global citizens in social, economic and political processes and exchanges that extend beyond borders and are often driven by a sense of duty and global justice, particularly to address the inequalities between the Global North and the Global South.

Critiques of cosmopolitan citizenship

There are many critiques of cosmopolitan citizenship. The most significant relates to the question of whether such a concept is even realizable. Republican theorists often argue that it is impossible for citizens to have meaningful allegiances outside the territorial state. There can be no citizenship without borders – their argument goes. Cosmopolitanism is actually bad for citizenship because it erodes 'the old home feeling' that helped produce a sense of mutual obligation and solidarity, in the absence of which there is no compulsion to exercise duties and no sense of responsibility (e.g. Miller 2000). This lack of public commitment and the thin conception of group identity will ultimately lead to the erosion of social and political bonds that are necessary to keep the political community together. Republican critics see in cosmopolitan citizenship similar problems to those they associate with liberal citizenship, but accentuated because of the increased distance between citizens and the lack of an external referent against which to create a sense of group identity and solidarity. In short, global citizenship is a less than viable proposition that empties the ideal of citizenship of its actual meaning (Bowden 2003). The extent to which these critics are correct in their general assessment or display a lack of imagination (after all national citizenship was 'impossible' until a few centuries ago) is something that only time will tell. In the meantime, and somewhat paradoxically, other critics argue that, far from being all-inclusive, cosmopolitan citizenship is in fact a very exclusive club – the club of the mobile, affluent classes with the means and the disposition to travel the world (e.g. Calhoun 2002). In other words, the universality of cosmopolitan citizenship is in fact a class particularity – defined by affluence and the culture of travelling that such affluence permits. The critique of cosmopolitan citizenship as a status exclusive to the global affluent is closely related with the final model of citizenship explored in this chapter: neoliberal citizenship.

Neoliberal citizenship: market rules

Neoliberal citizenship is the product of the neoliberal ascendancy that followed the collapse of the Bretton Woods system in the early 1970s and the coming to power of the 'New Right' in Britain and the United States in the 1980s. The neoliberal turn was driven, first and foremost, by neoliberal economics, embraced with particular fervour by Margaret Thatcher and Ronald Reagan (Harvey 2005). These leaders embarked upon a series of reforms to minimize state intervention in the economy that included a frontal attack against the welfare state, the privatization of industries and services, and the gradual deregulation of financial markets. This trend was also part of Deng Xiaoping's new economic policy in China, launched in 1978. The neoliberal paradigm has since been promoted across the world in the form of 'neoliberal globalization' (Kuisma 2008) by global institutions like the International Monetary Fund (IMF), the World Bank and the World Trade Organization (WTO).

The term neoliberalism suggests a return to liberalism. However, the implications of the prefix 'neo' go much deeper than a mere return to classical liberalism. Neoliberalism signals a radical shift from the state to the market as the locus of wealth production and distribution. But its impact goes way beyond the economy. The neoliberal paradigm applies market values and market rules to other aspects of life, from health, education and political participation, to family life and personal relations (Brown 2005). The extension of market rules to social and political relations has led some authors to argue that we are moving from being societies with a market economy to becoming market societies, particularly in the affluent West (Sandel 2012). In this context, it should not be too surprising to find out that neoliberalism poses the most profound challenge to traditional notions of citizenship. This challenge relates both to the 'who' and 'what' of citizenship (membership, rights and duties), and has materialized in the formulation of subsets of neoliberal citizenship, most notably corporate citizenship and consumer citizenship.

Neoliberal citizens: corporations and consumers

The neoliberal ascendancy has translated into the addition of a new member to the citizenship club: corporations. In recent years, corporations have emerged as powerful players in the sphere of citizenship. Their involvement derives in part from their influence over governments (and their policies). However, what makes corporations particularly significant is their status as legal persons – a status that has enabled them to claim the status of citizens in their own right. The inclusion of corporations as citizens represents a profound shift in the conception of citizenship, from the traditional dyad (individual–government) to the neoliberal triad (individual–government–corporation). Indeed, given the emphasis neoliberalism places on minimizing government and shifting relations to the market sphere, there is potential for a new, neoliberal dyad to emerge as central to the future of citizenship: individual–corporation.

The impact of neoliberalism has also been felt in the redefinition of the traditional citizen, the individual. Market theories of political exchange reduce the citizen to a 'consumer' or 'customer'. In essence, neoliberal citizenship shifts the focus from the individual as a political being (*zoon politikon*) towards the individual as an economic being (*homo economicus*). The extent and significance of this transformation are a matter of much debate. On the one hand, the consumer-citizen can be seen as a traditional citizen, but with additional power: consumer power. However, that shift can also be interpreted as the replacement of one type of member (the citizen) with another (the consumer), making membership and/or the exercise of citizenship contingent upon wealth and the capacity to consume, and ultimately transforming citizenship into consumption (Figure 2.5). The tension between consumption and citizenship is central to some of the most important works on environmental politics and economics, most notably Mark Sagoff's *The Economy of the Earth* (1988).

Depoliticizing citizenship: it's the economy, stupid

The neoliberal turn has also had a profound impact on the content of citizenship. Neoliberalism redraws the picture of rights, reinvigorating some classic rights (most notably, property rights), sidelining others (in particular, social rights), and creating or enabling the creation of new ones (mainly consumer rights). Neoliberalism also redraws the picture

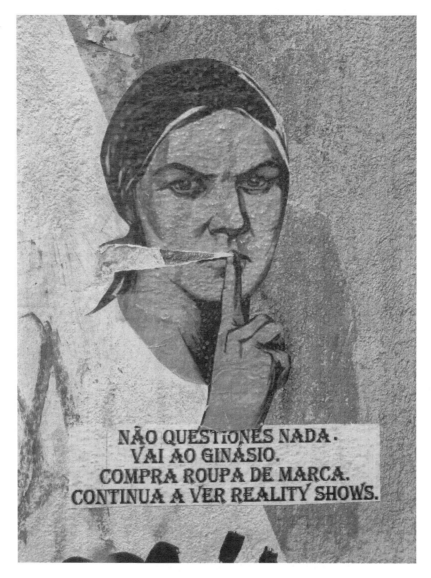

Figure 2.5 *'Don't question anything. Go to the gym. Buy designer clothes. Keep watching reality shows'. Street art in Porto, Portugal (2010).*

Source: Personal collection. Photographer: Benito Cao.

of duties, redefining some social rights as duties (particularly the right to work), shifting the emphasis from political to economic participation, and promoting the individualization of responsibility. Neoliberal citizenship also includes the rights and duties of its new member, corporations.

Traditionally, corporations have enjoyed commercial rights (as economic entities), but in recent times corporations have been able to claim and exercise civil and political rights – most notably, the right to free speech and the right to participate in political campaigns (in both cases, in the United States). Their duties are associated with corporate social responsibility (CSR), a concept we will explore at length in Chapter 6. In essence, the neoliberal portrait of rights and duties shifts citizenship from the formal political arena to the economic terrain and, to some extent, transforms citizenship from a political into an economic activity. This shift expresses itself in the two fundamental duties of the neoliberal citizen: work and consume.

Work: from social right to social duty

The most studied impact of neoliberalism on citizenship is the erosion of social rights that has accompanied the rolling back of the welfare state, and its transformation into the workfare state. The neoliberal common sense has been that social rights, in addition to being wasteful, inefficient and undermining individual freedoms, are creating inter-generational 'welfare dependency' (e.g. Dixon and Frolova 2011). Neoliberal theorists have advocated the privatization of social services (e.g. health, education, pensions) entrusted to the state for their provision as citizen rights, especially in the social and liberal democracies of the Global North. One of the most significant changes has been to the provision of income support for the unemployed. Indeed, employment has been transformed from a social right (the right to work) into a social duty (the duty to work). In the neoliberal model, the citizen has 'obligations to society and the family, which should be expressed through work' (Nash 2000: 197). However, this shift from a 'rights discourse' to a 'duties discourse' (Roche 1992) is *not* reflective of a shift from liberal to republican citizenship. The duties discourse shifts the emphasis from the political to the economic, from collective self-government to individual self-provision and self-regulation, and in doing so produces a new, hybrid model of citizenship: neoliberal citizenship. The neoliberal citizen, unlike the liberal citizen, is expected to be an active and dutiful citizen. But, unlike the republican citizen, the neoliberal citizen is not expected to be an active political participant (whose duty is to promote the public good, to improve the community) but an economic agent (whose principal duty is to work, so as not to burden the community).

Consume: vote with your wallet

The neoliberal ascendancy has been accompanied by the emergence and strengthening of consumer rights – the rights citizens have as consumers, when purchasing goods and services. These rights have resulted from the activism of consumers (and consumer groups) and have contributed to the protection of citizens when these operate (as consumer-citizens) in the marketplace. To be sure, consumer activism is not new, and neither are consumer rights (e.g. Glickman 2009). However, in the context of the neoliberal framework, consumption is becoming part of a deeper transformation to the meaning and practice of citizenship – one that encourages citizens to exercise their political agency, first and foremost, as consumers, voting with their wallet. This idea is rooted in classical market theory. Early in the nineteenth century, Austrian economist Frank Fetter wrote: 'The market is a democracy where every penny gives the right to vote' (quoted in Dickinson and Carsky 2005: 25). The force and popularity of this idea have increased greatly with the rise of neoliberal governance in the 1980s and 1990s.

There is now a broad consensus that consumers are key actors in the global political system, but there is still substantial debate about the extent to which consumption enhances the capacity to exercise citizenship in a neoliberal context (e.g. Soper 2004). Neoliberal theorists argue that consumption is a powerful tool through which citizens can influence public policy. In fact, studies on consumer movements and earlier forms of consumer boycotts show that, for a long time, consumption 'functioned as an alternative sphere of political action and inclusion for groups excluded from the formal body politic' (Trentmann 2007: 149). Currently, consumer activism often collapses the distinction between 'consumers' and 'citizens' suggesting that 'voting with your dollar' is a significant and perhaps more effective form of political participation, especially when it comes to certain policies, such as environmental ones (e.g. changing fishing practices by buying safely harvested tuna). The convergence of citizenship and consumption remains a matter of much debate (e.g. Soper and Trentmann 2008), and one of the areas that raises significant criticism regarding neoliberal citizenship.

Critiques of neoliberal citizenship

Much of the criticism of neoliberal citizenship relates to its close association with capitalism and resembles the critiques launched against liberal citizenship, particularly regarding inequality. In this sense, critics bemoan the undoing of the welfare state, which undermines the positive influence that social rights had in reducing social inequality (Shaver 2004). The shift from the citizen to the consumer, which explicitly positions money at the heart of citizenship, suggests that, if anything, inequality is now even more pronounced. Membership (who is included and who is excluded) and the exercise of citizenship (the capacity to participate) become contingent upon wealth and the capacity to consume. In this sense, Kaela Jubas asks: 'Can we buy our way out of the margins?' (2007: 249). Consumer citizenship effectively treats the poor as second-class citizens. In addition, critics argue that the power of voting with your wallet is often more illusory than real, especially in light of the increasing power of corporations (Korten 1995). In these critiques, the state is often portrayed as the enabler and protector of the interests of corporations ahead of the interests of citizens. Critics also argue that the discourse of consumer choice masks the coercive nature of neoliberal citizenship, which compels the citizen to produce and consume. You can choose your work (depending on your education, etc.) and what to buy (limited by your income) but you *must* work, and you *must* shop. And above all, you must choose. Individuals 'are not merely "free to choose", but *obliged to be free*, to understand and enact their lives in terms of choice' (Rose 1999: 87). Finally, those who view the political as the essential dimension of citizenship fear that neoliberal citizenship (with its emphasis on the economic dimension) might signal the end of citizenship as we know it.

Summary points

- The classical models of citizenship, republican and liberal, share a long history that goes back to ancient times, and they remain the dominant ones when it comes to frame debates and issues regarding contemporary citizenship.
- Republican and liberal citizenship share one fundamental trait: they both operate with a neat distinction of the private and the public, and associate the public with the sphere of citizenship. Historically, they have both been articulated in the context of bounded political communities. Otherwise, they differ in significant ways.
- Republican citizenship is a duties-based citizenship that privileges the collective, prioritizes the public good and demands from the citizen active political participation in the life of the community.

- Liberal citizenship is a rights-based citizenship that privileges the individual, prioritizes the private sphere and demands little from the citizen beyond law abidance.
- Pluralist theories challenge the universalism associated with mainstream theories, noting that certain groups have been and still are excluded from full citizenship because they are deemed to be 'different' (e.g. women, sexual minorities, indigenous peoples, ethnic minorities, etc.).
- Feminist theories challenge the public/private divide, expose the gendered character of citizenship, argue that the personal is political and reveal the centrality of time for the exercise of citizenship.
- Multi/culturalist theories expose the supposed neutrality of the modern liberal state regarding cultural matters, and advocate the recognition and granting of rights to cultural minorities.
- Cosmopolitan theorists challenge the inclusion/exclusion dynamic associated with the traditional conception of citizenship as a bounded political community, and argue for a global citizenship that entails the protection of human rights and shared global responsibilities.
- Neoliberal theorists shift the focus from the citizen to the consumer and from the state to the corporation (as agents of citizenship), and from politics to markets (as the sphere of citizenship). These shifts promote consumer and corporate citizenship and transform the citizen from a political being (*zoon politikon*) into an economic being (*homo economicus*).

Selected readings

Bowden, B. (2003) 'The Perils of Global Citizenship'. *Citizenship Studies*, 7(3): 349–362.

Ellison, N. (1997) 'Towards a New Social Politics: Citizenship and Reflexivity in Late Modernity'. *Sociology*, 31(4): 697–717.

Heater, D. (1999) *What Is Citizenship?* Cambridge: Polity Press.

Hindess, B. (2002) 'Neo-liberal Citizenship'. *Citizenship Studies*, 6(2): 127–143.

Isin, E. F. and B. S. Turner (eds.) (2002), *Handbook of Citizenship Studies*. London: Sage.

Jubas, K. (2007) 'Conceptual Con/fusion in Democratic Societies: Understandings and Limitations of Consumer-Citizenship'. *Journal of Consumer Culture*, 7(2): 231–254.

Kymlicka, W. (1995) *Multicultural Citizenship: A Liberal Theory of Minority Rights*. New York: Oxford University Press.

Linklater, A. (1998) 'Cosmopolitan Citizenship'. *Citizenship Studies*, 2(1): 23–41.

Lister, R. (1997) *Citizenship: Feminist Perspectives*. 2nd edn. New York: New York University Press.

Lister, R. (2007) 'Inclusive Citizenship: Realizing the Potential'. *Citizenship Studies*, 11(1): 49–61.

Online resources

Citizen/Soldier (uploaded 8 October 2009). Duration 3.34 mins. Music video by 3 Doors Down, with footage from The National Guard. Available at: www.youtube.com/watch?v=pgV6VUinDEA

Deutsche Bank: Corporate Citizenship (published on 11 October 2012). Duration 1.53 mins. Promotional video presenting Deutsche Bank as a good corporate citizen. Available at: www.youtube.com/watch?v=YPPFjoeykmM

How Does One Become a Global Citizen? (published on 21 May 2012). Duration 17.34 mins. Tanja Schulze's talk on global citizenship at TEDxBMS. Available at: www.youtube.com/watch?v=A4XF8GCXYtM

La ciudadanía desde el feminismo anarquista (published on 21 August 2009). Duration 47.38 mins. Keynote address by Ximena Castilla on anarcho-feminist citizenship. [In Spanish]. Available at: http://hidvl.nyu.edu/video/003475792.html

Neoliberalism vs Democracy (uploaded 10 June 2012). Duration 9.38 mins. Excerpt from Richard Brouillette's documentary *Encirclement* (2008). Available at: www.youtube.com/watch?v=g-wHipDwvgk

Oxfam Education: What is Global Citizenship? The meaning and value of global citizenship, according to Oxfam. Available at: www.oxfam.org.uk/education/global-citizenship/what-is-global-citizenship

United Breaks Guitars (uploaded 6 July 2009). Duration 4.36 mins. Protest song by Canadian musician Dave Carroll and his band, Sons of Maxwell. Available at: www.youtube.com/watch?v=5YGc4zOqozo

What Is CitizenCard? (uploaded 27 February 2011). Duration 2.57 mins. Video about the CitizenCard, the UK's leading proof of age card ID. Available at: www.youtube.com/watch?v=aNsjJ3dK9CQ. See also the website: www.citizencard.com/

Student activity

Explore your citizenship. Select one or two items that you associate with your citizenship (e.g. passport, social security card, voter registration form, tax return, credit card number, or any other personal document or artifact that you associate with your sense of citizenship), and try to establish which aspects of citizenship that item relates to or illustrates. Does that item say something about your membership, rights and/or duties? Does it say something about your identity as a citizen? If so, what does it say? Does it relate to your local, national or global identity?

Having identified how the item relates to these elements, try to relate them to different conceptions of theories. What theories do you think better relate to that item, e.g. liberal, republican, feminist, multicultural, cosmopolitan or neoliberal? You can watch the videos listed above to get ideas for this activity. They include, amongst others, examples of republican citizenship and consumer citizenship.

Discussion questions

- Which of the two classical theories appeals the most to you? Do you see yourself mainly or primarily as a republican or a liberal citizen?
- Contrast the symbolism of the titles given to the foremost documents of modern and global citizenship, respectively: Declaration of the Rights of Man and the Citizen (1789) and Universal Declaration of Human Rights (1948). How significant, if at all, do you find the different vocabulary?
- Do you consider yourself a global citizen? If so, why? If not, why not? Take a look at Oxfam's description of global citizenship (see online resources). Does that make you change your views?
- Do you see yourself as a consumer-citizen? If so, why? If not, why not? Do you think that consumer citizenship enhances or undermines democratic citizenship?

3 Theorizing environmental citizenship

> The severity of the world wide environmental crisis demands the 'greening' of our conception of citizenship.
>
> David Chamberlain (2007)

The increasing preoccupation with environmental sustainability and the injection of green values into political analyses of 'how we ought to live' pose significant challenges to existing conceptions of citizenship. There is significant debate over whether these concerns and values can be accommodated by the traditional language and the current theories of citizenship or whether new and distinctive conceptions are required. This has led to the formulation of a wide range of competing theories of environmental citizenship since the term gained currency in the early 1990s. Many attempt to address these challenges without abandoning traditional theories and formulations. Others stretch the concept significantly, sometimes even radically, producing alternative conceptions of citizenship. In this chapter, we will scope and explore both approaches. The first part explores models of environmental citizenship that result from adjustments to established theories. The second looks at theories that propose significant alterations to the classic architecture of citizenship, some of which signal a radical departure from traditional conceptions of citizenship. In recognition of its widespread influence, a whole section is dedicated to the notion of 'ecological citizenship' theorized by Andrew Dobson. This will be followed by a brief account of alternative and emerging theories, approaches and reflections, many of which result from direct and explicit engagement with Dobson's work. But we begin by exploring how the classical and most popular conceptions of citizenship (i.e. liberal, republican and cosmopolitan) have taken up the 'green' challenge.

Liberal environmental citizenship

The greening of citizenship poses important challenges to its traditional conceptions, but perhaps none more so than to the liberal conception of citizenship. The integration of the environment into citizenship questions the extent to which a model of citizenship centred on the individual (a thoroughly anthropocentric and individualistic conception of citizenship) can handle the structural challenges associated with environmental sustainability and accommodate nature into its formulation of citizenship. Yet, liberal thinkers have produced several formulations of green citizenship, suggesting that liberalism can accommodate concerns with sustainability, and even ecological values. In this section we explore the greening of liberal citizenship, with particular attention to the most influential articulation of liberal environmental citizenship, the work of Derek Bell.

Liberal citizenship, as we have seen, is essentially a rights-based notion of citizenship that aims to maximize individual freedom. Liberalism conceives citizenship as a status that grants individuals legal protection and allows them to pursue their private interests. The liberal model is committed to valuing pluralism and deploys the rule of law to guarantee the protection of rights, including everyone's right to pursue their own conception of the good. Liberalism aims to protect the right of each individual to choose the values and morals they wish to live by, the conception of the good life they desire, limited only by the fundamental principle of doing no harm to others. In other words, liberal citizenship prioritizes legality (matters of justice) over morality (matters of virtue). So, how does the environment impact on and fit into this articulation of citizenship? What are the liberal responses to the challenges posed by concerns with environmental sustainability?

One of the first authors to address the green challenge was Marcel Wissenburg (1998). His work seeks to identify the level of compatibility between liberalism and green thinking, and to assess how green liberal policies can become without violating fundamental commitments of liberal political thought. Wissenburg notes that liberals value sustainability as something important, but they do not prescribe how sustainability should be achieved or what a sustainable world should look like. Individuals should determine that. He argues that

> liberals must reject the deeper green positions of ecocentrism and the
> intrinsic value of nature on the following grounds: the first violates

the most basic premise of liberal thought, the freedom to construct
one's own life plans; and the second is confused by its failure to
recognize that all value stems from a subject.

(Gabrielson 2008: 432)

Wissenburg notes that liberals may be obliged to act in a manner
consistent with deep green demands, but the theory does not exclude 'the
possibility of turning the planet into a giant steel-grey Manhattan' (1998:
208). His commitment to value pluralism fails to acknowledge the extent
to which a particular conception of the natural world (nature as property)
is entrenched in liberal thought – an issue taken up by Bell (2005).

The most cited and influential formulation of liberal environmental
citizenship is arguably that of Derek Bell. In his view, the main problem
for a liberal approach to environmental citizenship is its conception of
nature as property. Bell argues that liberalism must replace this view with
one that sees the environment as the means of human survival, noting
that 'the liberal commitment to environmental sustainability should
trump property rights to (part of) the physical environment' (2005: 183).
In other words, the liberal concern about current and future generations
being able to meet their basic physical needs should be grounded in
a conception of nature as 'provider of basic needs' (Bell 2005: 183).
He also notes that the liberal emphasis on the freedom to choose and
pursue our particular conceptions of the good life means that individuals
should be able to live a life in which they are connected to the physical
environment, if they so wish. In this view, damage to the environment
curtails the choices of those who desire this particular way of life and
impacts on the 'reasonable pluralism' that is central to liberalism. But
in keeping with the liberal tradition, Bell also notes that we should
regard the environment as 'a subject about which there is reasonable
disagreement' (2005: 190), rather than something that (as it is might be
for deep ecologists, for example) has intrinsic and undisputed value of
one type or another. Liberal citizens thus conceived are not citizens of the
environment but 'citizens of *an environment*' (Bell 2005: 186).

Bell then sets out the kinds of rights and duties compatible with this
conception of liberal environmental citizenship. The list of rights includes
substantive rights (specific goods, such as the right to clean air and clean
water), procedural rights (the right to participate in the decision-making
process over environmental policy), and personal rights (the right of
individuals to make green choices in the way they go about their lives).
The exercise of these rights is always a matter of choice. This model
of citizenship does not require individuals to participate in activities

seeking to protect the environment. Liberal environmental citizens
have the right *not* to be greens. The fundamental duties accompanying
these rights are: the duty to obey just laws (and thus the right to resist
unjust ones) and the duty to promote just arrangements. These duties
are limited by a cost proviso that establishes that we should only seek to
further just arrangements when this can be done without too much cost
to ourselves (Bell 2005: 184). Bell argues that no one can be asked to do
things that are too costly, that is, things that would cause more harm (in
terms of freedom) than good (in terms of sustainability). In other words,
environmental demands on citizens must be reasonable.

The importance attached to personal freedom and the rule of law in
liberal conceptions of citizenship finds another interesting articulation
in the work of Bruce Pardy. Pardy argues that 'liberty does not mean
the absence of rules' but 'rules that protect one from interference from
others' (2005: 33). In other words, 'liberty requires mutual restrictions
in the form of protection by the state – from the state, and from other
people' (Pardy 2005: 33). Pardy resorts to the basic proposition of the
law of torts to make his argument – the notion that 'you may do as you
please, but you may not in the exercise of your liberties cause harm to
others' (2005: 33). He then proceeds to note that, given that ecosystem
limits cannot be expanded, individuals interfere with and harm the
ecological rights of fellow citizens when they use more than their share
of ecological resilience. Thus, the 'right of ecological citizens to be free
from interference and harm from others would consist of the right to
undiminished ecosystem services – their full slice of pie – since these
services are indispensable to human survival, biologically, economically,
and in other ways' (Pardy 2005: 34). In other words:

> Individual citizens should conform to the limits of the ecosystem
> within which they find themselves – not because it is morally right,
> but because it is the (legal) right of fellow citizens to be free from the
> ecological interference of others.
>
> (Pardy 2005: 34)

The liberal value of reasonableness is central to the accommodation of
green thinking in the liberal conception of environmental citizenship
formulated by Simon Hailwood. His point of departure is the notion
that liberal citizens must give reasons for their political demands, not
just state preferences or make threats. Moreover, those reasons must be
public reasons, capable of being understood and accepted by people of
different faiths and cultures, in ways that are consistent with their status
as free and equal citizens. Hailwood tries to make compatible this view of

reasonableness (drawn from the liberal tradition) with a non-instrumental view of the natural world (reflecting an ecological approach to green citizenship). The infusion of ecologism into a liberal framework leads him to argue the duty to protect 'nature as other' – and not just because its preservation is required for the sake of enabling a future diversity of good lives (Hailwood 2005). In other words, nature must be protected for its own sake and value (as any other subject, that is, as any given 'other') and not simply because of the liberal ideal of not foreclosing particular human opportunities.

In general, liberal environmental citizenship retains the centrality of the individual and individual rights, protected by the rule of law. This is a form of citizenship exercised by a rights-bearing citizen that entails, above all, the addition of several substantive and procedural rights. In addition to the classical liberal rights, this citizen has the right to an environment that can provide the basic human needs of potable water and clean air (linked to the basic right to life), the enjoyment of nature (as individual pleasure) and a balanced ecosystem (as individual right). These rights come with the basic liberal duty of compliance with the law, in this case with environmental laws, provided those laws have been determined by argument (public reason) and agreement (democratic process) and are considered fair and just (not unduly onerous on the individual). Liberal environmental citizens retain the right to resist specific policies if they demand excessive sacrifices or are deemed to be unjust or unduly coercive. The level of participation is decided by each and every individual. Citizens are free to engage in environmental actions as much as they wish, but are under no obligation to do anything else other than complying with reasonable legislation. In essence, liberal environmental citizenship relies on individual choice and voluntary actions.

Critics consider these attempts to bring the environment into the framework of citizenship entirely insufficient to address the environmental challenges facing humanity. The emphasis on individual choice and voluntarism fails to generate the sense of urgency, the level of commitment and the amount of action required to tackle the environmental problems we face and that will be inherited by future generations. Moreover, the focus on the individual gives insufficient consideration to structural issues and concerns that many consider central to the current environmental crisis. Indeed, for many, particularly for deep greens, the liberal model embodies the instrumental rationality, anthropocentrism and economic expansionism largely responsible for environmental degradation in the first place (Naess 1989). Some critics find a better answer to our environmental predicament in the greening of the other traditional theory of citizenship: republican citizenship.

Republican environmental citizenship

The challenges posed to theories of citizenship by concerns with sustainability explain in part the recent revival of republican citizenship. Indeed, to the extent that much of the talk about environmental citizenship comes in the form of references to values, duties and responsibilities, republican citizenship is very much in tune with attempts to address our environmental predicament. Having said that, green issues also pose significant challenges to republican citizenship, in particular to its traditional and strong association with bounded political communities. Some theorists have tried to address the challenge within the republican tradition. In this section we take a look at notions of republican environmental citizenship, with particular attention to the most influential of them, the work of John Barry.

Republican citizenship, as we have seen, is essentially a duties-based conception of citizenship that aims to protect the common good. Republican theorists conceive of citizenship as action, giving primacy to participation in the political life of the community. The republican model aims to guarantee citizen rights through the protection of the collective and the public interest. The emphasis is on the promotion of virtues and the creation of norms essential to the survival of the political community and the protection of the public good. In other words, republican citizenship prioritizes the moral dimension (what is good) over the legal dimension (what is right). So how does the environment impact on and fit into this articulation of citizenship? Can republican citizenship accommodate environmental concerns in ways that will contribute to a sustainable world? What are the republican responses to the challenges posed by concerns with environmental sustainability?

The most influential conception of republican environmental citizenship to date is that of John Barry. Barry argues that the characteristics typically associated with classical republicanism (virtue, duty, obligation and public service) are best suited to the promotion of 'sustainability citizenship' (2006). This conception of citizenship aims at repairing not only environmental problems but also their roots and causes, pursuing structural changes as well as 'lifestyle' ones. Barry argues that institutional reforms are insufficient and must be 'supplemented with changes in general behavior (weak green citizenship) and values and practices (strong green citizenship)' (2002: 147). He also makes the point that 'if one accepts the argument for sustainability, one does not just have the *right* to demand changes to create a more sustainable society but one also has the *obligation* to do so' (Barry 2006: 33). This approach to sustainability demands a deep commitment to environmental action and participation, and creates a 'politics of obligation' (Smith 1998).

The sense of urgency, combined with the belief that change will not be substantial and sustained if it is not underpinned by values, has revived the interest in fostering civic virtue. Significantly, some of the virtues often associated with republicanism, such as frugality and disdain for luxury, dovetail neatly with popular environmental attitudes towards consumption. This suggests – and Barry hopes – that tapping into the republican tradition could help environmentalists promote 'mindful as opposed to mindless consumption, and to seek a balance between the extremes of excessive consumption and no consumption/poverty' (Barry 2006: 38). In general, Barry advocates a conception of citizenship based upon intellectual virtues (a basic ecological literacy grounded upon scientific knowledge) and moral virtues such as self-reliance, self-restraint, prudence, foresight and the consideration of non-citizen interests. But how do we become virtuous? How do we become green citizens?

Barry notes that, in the republican tradition, 'citizenship is something that has to be learned rather than something that comes naturally to members of society' (2006: 27). This means that citizenship in general and green citizenship in particular can be taught and encouraged, but also forgotten. Barry attributes the failure to move decisively towards sustainability to the tendency of citizens to 'forget their duty or lapse into self-regarding interests and pursuits' (2006: 27). He is critical of passive or voluntary notions of green citizenship and emphasizes the need to cultivate virtue through practice – including participation in decision-making on environmental issues and 'collective ecological management' (Barry 1999: 5). In this context, the republican notion of public service can offer an antidote to such inertia and provide a means to cultivate more environmentally inclined citizens (Figure 3.1). Yet, the fact that teaching green values and virtues, and general environmental awareness, does not necessarily translate into action means that the state might need to step in and become actively involved – using coercion if necessary. Barry proposes a form of 'compulsory sustainability service' (enforced by the state) that would have citizens cleaning up polluted rivers, working

Figure 3.1 *'Your Planet Needs YOU to Recycle'*

Source: Poster courtesy of Stacey Griffiths and Carmel College, UK.

in community-based recycling schemes, or participating in campaigns of environmental education (2006: 28–32).

Significantly, Barry presents sustainability citizenship as a form of 'resistance citizenship' (2006: 32). His claim is based on the fact that this model of citizenship exists in the midst of and challenges unsustainable development. Sustainable citizenship, he argues, requires 'maintenance' activities but also 'corrective' or 'oppositional' practices that challenge the underlying causes of unsustainable development (Barry 2006: 32). In this sense, Barry's concept of sustainability citizenship should not be associated with more obedient or submissive forms of republican citizenship. This point becomes particularly clear when he notes that: 'In the struggle for more sustainable, just, and democratic societies, we need civil disobedience before obedience, and, more than ever, we need critical citizens and not just law-abiding ones' (Barry 2006: 40). Sustainability citizenship ultimately translates into a call for ecological action, inspired by the notion that 'resistance is fertile' (Barry 2006).

The virtue approach to green citizenship is also present in the work of James Connelly. Connelly argues that virtues (defined as continuous and reliable dispositions) are central to green citizenship. He shares with Barry the view that virtues are learned and that virtue cannot be theorized into being, but that 'one must participate in a practice to discover its internal goods and goals' (Connelly 2006: 67). Connelly also cites frugality and asceticism as essential eco-virtues that need to be promoted, but adds that care and compassion should have a role to play in our understanding of green citizenship. His list of eco-virtues includes: frugality, care, patience, righteous indignation, accountability, asceticism, commitment, compassion, concern and cooperation. Connelly also believes that the promotion of environmental virtues demands state action, possibly through legislative carrots and sticks, and concludes that: 'a sustainable society will continue to require law, regulation, and economic incentives whose presence serves as a moral indicator of values and goals' (2006: 71).

Some republican authors take up a more ecological approach. Thus, for example, Patrick Curry emphasizes the common good, including the natural environment, and argues that its protection can only be maintained through the exercise of active citizenship and the shared commitment to a set of practices that uphold the public good against the corrupting influence of self-interest. In his words: 'communities are only maintained by certain practices, in default of which they disintegrate'

(Curry 2000: 1068). But Curry extends the community beyond humans, proposing 'an ecological republicanism, in which the natural world, without "determining" specific outcomes, would once again provide the context of human political, social and ethical deliberation' (2000: 1068). He views humans as members of a republic of life, and calls on us to maintain and encourage the continued diversity of plants, animals and their habitats that make up that natural-political community. Similarly, Deane Curtin argues that 'the moral orbit of citizenship' must 'be extended to ecological communities' and articulates the notion of a 'more-than-human community' (2002: 293).

In general, the republican model accommodates environmental concerns through its emphasis on duties and obligations – which under notions of green citizenship extend to future citizens and, depending on the author, can also extend to nature itself. The essence of republican environmental citizenship is the primacy of the common good (again, one that sometimes includes nature as part of the community). This conception of green citizenship entails the addition of environmental virtues and duties (e.g. mindful consumption) to the classic notion of republican citizenship. In addition, this model of citizenship calls for the individual, at times directed or coerced by the state, to take a more active role in political life. In other words, republican environmental citizenship is exercised by virtuous citizens who can be compelled by the state to act responsibly if they fail to do so voluntarily. The greening of republican citizenship is largely an attempt to encourage and create an identity and mode of thinking and acting, and ultimately character traits and dispositions that accord with the standards and aims of ecological stewardship.

The critics take aim, first and foremost, at the coercive and demanding potential of this model of environmental citizenship, one that can lead to the imposition of specific notions of the common good by influential groups and powerful interests. The inclusion of nature in some formulations of republican environmental citizenship is also a significant source of criticism, both from within and outside the republican tradition. The main point of contention is the extent to which the natural community can be accommodated into a theory of citizenship that emphasizes the political and has always been associated with human, bounded political communities. In other words, can the notion of a natural community be integrated into the notion of political community? Or should the former replace the latter as the constitutive community of citizenship? And if so, can we still call this citizenship, republican or otherwise? In addition, critics note that republican citizenship is still very much attached to the national space, whilst environmental matters often, and increasingly, demand thinking beyond the nation-state, or any bounded political community for that matter. This challenge is what has brought to the fore of environmental citizenship the model to which we now turn: cosmopolitan citizenship.

Cosmopolitan environmental citizenship

Critical environmental issues, such as ozone depletion, nuclear waste and climate change, transcend national borders and demand transnational solutions and cooperation. This has made cosmopolitan citizenship an attractive and even necessary proposition. Indeed, the fact that these issues (and environmental issues in general) do not respect the artificial boundaries of nation-states has even led some to argue that 'citizens sensitive to the environment must be global citizens' (Nash 2000: 212). The connection between the environment and cosmopolitanism has several dimensions. On the one hand, environmental risks are an incubator of cosmopolitan emotions and sensibilities. Climate change, in particular, releases a 'cosmopolitan momentum' that transcends national borders, creating a greater sense of interconnection and interdependence (Beck 2010). On the other hand, because such risks are global in scope, they can only be solved by means of global cooperation between countries and citizens (Linklater 2007). In short, environmental risks require that citizenship finds cosmopolitan forms of expression (Stevenson 2002).

The work most commonly associated with the conception of cosmopolitan environmental citizenship is that of Elizabeth Jelin. Her work does not propose an alternative model of citizenship, but focuses instead on the main implications brought about by the global nature of most environmental issues. Jelin begins by pointing out that the introduction of environmental issues in considerations about citizenship implies a paradigmatic shift in the conception of citizenship towards a global framework. She notes, with a clear sense of historical perspective, that

> Although the ideas about citizenship and rights have been grounded in the notion of the modern nation-state, there is no intrinsic necessity that this be so: the public sphere might be 'smaller' or 'larger' than the state, or may even be different.
>
> (Jelin 2000: 53)

She clearly settles for the 'larger' political community, the cosmopolis.

Jelin refuses to suggest a specific list of environmental rights and duties. She presents the right to have rights as 'the motor of historical change' and leaves open the question of which rights and duties should emerge from the energy generated by that motor. She even refuses to produce a 'basic agenda', calling instead for a process of 'debate, dialogue and struggle' that is more flexible, more dynamic and more varied than has been to date (Jelin 2000: 59). Yet, she is keen to point out that the recognition of the contingent nature of struggles and demands should not imply abandoning ideals and utopias, but merely taking on and recognizing that there are no absolute truths. The focus, she argues, should be on the materialization of a particular set of generic ideals, which she identifies as: 'the elimination of suffering and oppression, the promotion of solidarity and a sense of responsibility for others, the concern for the quality of life and for sustainability' (Jelin 2000: 59). She also leaves open the question of 'the status of nature', that is, whether nature has (or should have) rights, and, if so, how to protect them.

The cosmopolitan approach to environmental citizenship has become increasingly popular, but not without its fair share of criticism. Most critics take aim at its political space, noting that shifting the space of environmental citizenship to the global can distract from the significance of the local (the space where the environment is experienced) and the national (the space that still dominates the articulation and implementation of environmental policies). Some critics point out that there is no need to give up the national level of political representation in order to address environmental questions, not even transnational ones. Not only can environmental issues be negotiated between nation-states but, in fact, the nation-state remains the only agent with the capacity to enforce environmental justice and administer environmental rights. Others position their critique from a defence of the local as the ideal space for the expression of environmental citizenship. These critics take aim at the popular saying 'think globally, act locally', noting that when local movements move into global arenas, they become minor players in a much bigger game, 'a footnote to a set of global agreements' that effectively ignores local concerns (Esteva and Prakash 1994: 162). Their proposal is to act and think locally – a position often inspired by the work of Murray Bookchin (see Box 3.1).

Box 3.1

The ecological anarchism of Murray Bookchin

Murray Bookchin was an influential American thinker and political activist who fused anarchism and ecology in an original theory called social ecology, which can be described as a form of eco-anarchism (Figure 3.2). Anarchists reject imposed authority, hierarchy and domination and seek to 'establish a decentralized and self-regulating society consisting of a federation of voluntary associations of free and equal individuals' (Marshall 2008: 3). Bookchin argued that the domination of nature by humans has its roots in the patterns of hierarchy and domination in human societies (of human by human). Therefore, all ecological problems would be resolved following an anarchist revolution. In his view, the non-hierarchical sensibility generated by the abolition of hierarchies amongst humans would naturally extend itself to the nonhuman world.

Bookchin is important for being one of the few environmental theorists to discuss the role of cities as part of the broader ecological project. Bookchin called for the dissolution of the state and a decentralized politics in which people organized themselves in direct democracies at the local level, without any representation or mediation, and formed loosely connected networks of municipal confederations or eco-communities. This vision was based on the idea that republican mutual obligation and direct democracy could not be demanded and enacted on a polity as large as the nation-state. Instead, he advocated for political rule and citizenship in entities no larger than the small town. He called this model of stateless social and political order 'libertarian municipalism' (Bookchin 1995). His vision demanded a return to a more classical type of local citizenship, as represented by Athenian democracy, which implies a community small enough that people are known to each other, regular citizen assemblies and a high degree of enlightened participation in political affairs. This implies the creation of local political communities guided by reciprocity, a sense of public service, cooperation and sharing. Bookchin concluded that: 'the recovery of a classical concept of politics and citizenship is not only a precondition for a free society; it is also a precondition for our survival as a species' (1995: 245).

Figure 3.2 *'Green Anarchism'*

Source: Wikimedia Commons.
Author: Riccardo Freeman.

The cosmopolitan approach to environmental citizenship is also criticized for acquiescing in and even enabling global environmental inequalities. The process of globalization, argue its critics, takes place in a context of asymmetrical capacities between countries (and citizens) to frame and influence global agendas. There is no 'diffusion' of ideas, goods, information, capital and people but their 'transfusion' from the Global North to the Global South (Dobson 2003: 15). In other words, some countries can only act locally whilst others act locally and globally. This critique relates to the tendency of global notions of the political to downplay the inequality between the Global North and the Global South. This critique informs, for example, the principle of a 'common but differentiated responsibility' regarding climate change – a principle that can be easily hidden behind, and is often absent from practical manifestations of cosmopolitan environmental citizenship, such as the Earth Charter.

In general terms, cosmopolitan environmental citizenship only goes beyond liberal and republican formulations in its embrace of the planet as the constitutive political community. The picture of environmental rights and duties that emerges from cosmopolitan models of environmental citizenship is relatively straightforward, and resembles the extensions of environmental rights and duties embraced by liberal and republican conceptions of green citizenship, respectively. The rights and duties contemplated tend to be expressed in the language of human rights and global duties – but are otherwise similar packages. Thus, in the same way that critics questioned the capacity of green articulations of liberal and republican citizenship to deliver sustainability, not everyone is convinced that the cosmopolitan approach is adequate or sufficient to address our environmental predicament. To address these limitations, other theorists have developed or suggested more ambitious and justice-based approaches. The most influential of these proposals is found in the work of Andrew Dobson.

Ecological citizenship: the work of Andrew Dobson

The most influential and systematic contribution to the field of environmental citizenship is the work of British political theorist Andrew Dobson. That influence derives from his conception of ecological citizenship, developed in his seminal text *Citizenship and the Environment* (2003). Dobson postulates a significant difference between environmental and ecological citizenship. He associates the first term

Figure 3.3 *Andrew Dobson's* Citizenship and the Environment *(2003)*

Source: Book cover image courtesy of Oxford University Press.

(environmental citizenship) to reformist articulations compatible
with conventional theories of citizenship (such as the ones we have
explored thus far in this chapter), and reserves the second (ecological
citizenship) for his distinctive formulation of green citizenship, drawn
from ecological thought. However, his ecologism is not of the 'deep
ecology' variety, with its associated ecocentrism. Dobson argues that
ecocentrism (in the context of ecological citizenship) can only be
metaphorical, since all conceptions of citizenship are by definition
human conceptions, based on what we (humans) value – even if what
we value happens to be nature (animals, plants, minerals). Yet, despite
this self-proclaimed anthropocentrism, his conception of ecological
citizenship is still quite radical compared with conventional articulations
of citizenship.

The model

Dobson argues that conventional notions of environmental citizenship
(i.e. those anchored in the classical citizenship traditions) fail to provide
a good enough account of what is required both to create a sustainable
world as well as one where matters of justice (which he takes to be central
to environmental politics) are given sufficient weight. Environmental
citizenship, he notes, typically takes the form of liberal citizenship
with a green tinge, mainly in the form of additional environmental
rights. Dobson associates this articulation of citizenship with reformist
attitudes and 'light green' politics that result in 'shallow' adjustments
that can fit neatly within mainstream notions of citizenship, particularly
market-friendly ones. He also finds problematic the continuing
association of liberal citizenship with the nation-state and the exclusive
identification of citizenship with the public sphere – two issues that also
reveal republican citizenship as ill-suited to respond to the demands of a
model of citizenship geared towards sustainability.

Dobson finds in cosmopolitanism a better answer to the challenge
of sustainability, agreeing with the notion that citizenship must
operate 'outside the realms of activity most normally associated
with contemporary citizenship: the nation-state' (2003: 97). He
rejects, however, the cosmopolitan association of globalization with
interdependence and interconnectedness amongst peoples and nations,
identified by figures such as David Held. He contrasts this idealistic
and idealized portrait with that of authors like Vandana Shiva, who
note the significant asymmetries operating at the heart of globalization,

particularly between the Global North and the Global South. Dobson takes the reproduction of daily life in an unequal and asymmetrically globalizing world as the context from where the political space of obligation for ecological citizenship emerges. In this space, obligations are not owed equally (by everyone to everyone), but asymmetrically (by some to others). This political space of asymmetrical obligation is very different from the formally neutral and egalitarian cosmopolitanism of global initiatives such as the Earth Charter.

Dobson also takes issue with the cosmopolitan conception of political community. He finds the bond of our common humanity too 'thin' to generate the level of political motivation required to mobilize people into significant environmental actions, and argues that the required political space of obligation that can result in citizen action will only result from developing a 'thick' cosmopolitanism – not in terms of identity but in terms of materiality (Dobson 2006a). In other words, the obligations of citizenship must stem from the systematic ecological injustice that has accumulated over time and defines much of our current environmental predicament. Here, the affluent become politically obligated to the poor for damaging the environment and, thereby, physically harming the poor. The general idea is that 'the patterns and effects of globalization have given rise to a series of material conditions within which the idea of transnational citizenship obligations can make sense' (Dobson 2003: 127).

Dobson anchors his model of ecological citizenship on the primary virtue of justice and the secondary virtues of care and compassion for distant others, both in time (future generations) and in space (around the world). His conception of ecological citizenship rejects conventional constructions of citizen obligations, associated with the artifice of the social contract (liberal and republican citizenship) or the moral commitment to our common humanity (cosmopolitan citizenship). Instead, the obligations of the ecological citizen are based on the obligation to correct the ecological injustices inherent in material relationships, reflected in the different sizes of our ecological footprints. The ecological footprint refers to the amount of ecological space that individuals, organizations and communities take up, in terms of resources used and burdens on the environment, that is, the amount of land required to sustain a particular lifestyle. Since the total ecological space is limited, this demands that we all take care that 'any ecological footprint with which we are associated (our own, that of our home, workplace and so on) is not so unjustifiably large as to cause harm to other people' (Dobson 2012: 521).

The ecological footprint replaces the nation-state, supranational entities and the cosmopolis as the political space of ecological citizenship. This metaphorical space unites people around the world and produces environmental obligations or entitlements depending on the size of each footprint. The inequitable distribution of ecological footprints creates obligations from those with excess footprints ('debtors') to those with insufficient resources ('creditors'). The obligations extend to future generations and to those who currently take up less than their share of ecological space – that is, those who tread lightly on the Earth. This generates a sense of 'causal responsibility' that connects everyone with everyone else and with everything, and allows Dobson to claim that 'the community created by the material relations of cause and effect in the guise of the ecological footprint is a political community' (2006b: 447).

The ecological footprint refers to the daily actions that have an environmental impact, which in practice extend to virtually everything, including energy consumption at home, the transport we use, the food we eat and the goods we purchase. In other words, ecological citizenship demands the recognition of private actions, such as composting, conserving energy, shopping and eating habits, which have long been considered nonpolitical activities (because they are private life activities), as deeply political, due to their public consequences. This politicization of the private sphere mirrors the feminist challenge to the public/private divide and evokes the feminist notion that 'the personal is political'.

In short, Dobson's conception of ecological citizenship draws on cosmopolitanism (the notion that we are all members of a global political community), republicanism (the notion that the common good is utterly dependent on the community, and maintained by practices and duties of active citizenship) and feminism (the notion that politics is not confined to the public sphere but intersects with the private sphere and the personal). Ecological citizenship, thus understood, breaks with conventional conceptions of citizenship in four important ways: (1) it is non-reciprocal and asymmetrical (rather than contractual and egalitarian); (2) it operates in the public and the private sphere (rather than only in the public sphere); (3) it emphasizes so-called 'feminine' virtues (e.g. care and compassion); and (4) it is non-territorial (rather than territorially bounded). This suggests a radical break with the traditional architecture of citizenship – although not everyone is convinced of such claims or of the merits of such a radical proposal.

The critiques

Dobson's work has received plenty of attention, and criticism. The strongest critique has come arguably from Tim Hayward. Hayward questions the claim that ecological citizenship is a new form of citizenship. Instead, he argues, ecological citizenship should be understood as giving distinctive substance to a conventional understanding of citizenship. He calls it 'new wine' in an 'old bottle' (Hayward 2006: 441). Hayward also takes issue with the definition of the ecological footprint as the political space of ecological citizenship. He calls the definition of the ecological footprint as a political community a 'category mistake' (Hayward 2006: 438). Hayward notes that the political community generated by the ecological footprint not only is not 'bounded' in any conventional way but it does not even have a determinate membership. Who belongs? In principle, everyone would belong, either as ecological debtor, creditor or equal. However, since Dobson identifies the ecological citizen as one who is bound by the obligations that arise from an excessive ecological footprint, this means that only those responsible for harm are ecological citizens, effectively casting the 'creditors' as 'victims' in the role of 'moral patients' (Hayward 2006: 445). How can 'creditors' influence and participate in this model of citizenship if they are, or at least seem to be, excluded (as active citizens) by definition?

One of the most intriguing critiques relates to the sidelining of justice – although this critique extends to the field of environmental citizenship in general, and not only to Dobson's work. Those who support this critique view the language of citizenship as a step back in the political struggle for global environmental justice (e.g. Agyeman and Evans 2006). If justice is the objective, they argue, better to stick with the language of justice than subsume or dilute justice into the language of citizenship. In a similar vein, others argue that a radical departure from some of the essential elements of citizenship (e.g. reciprocity and the private/public divide) can damage the concept of citizenship whilst adding nothing to the fight over environmental justice (Mason 2009). Others, however, see the value in deploying the language of citizenship and suggest that 'environmental justice can be read *in terms of* a politics of citizenship' (Latta 2007b: 386).

The other major criticism levelled at the concept of ecological citizenship relates to the lack of sufficient attention to the structures within which individual agents operate. The focus on individual action sidelines structural constraints that many believe are at the heart of the ecological crisis. In addition, the emphasis on individual responsibility and the voluntary nature of ecological citizenship dovetail (or at least resemble) neoliberal thinking. This is a somewhat paradoxical twist given that Dobson formulates ecological citizenship as an alternative to the market-friendly notion of environmental citizenship. Yet, as some critics have noted, individualist conceptions of ecological citizenship can easily be co-opted by government and state agencies who wish to step back from environmental regulation (MacGregor 2006b). This line of criticism also points out the dominance of lifestyle politics associated with ecological citizenship, as well as with the ecological footprint (an issue we will explore in Chapter 7).

Dobson's work has also been criticized for a latent elitism that derives from the focus on individual acts and lifestyle politics at the expense of collective action and survival politics that are more prevalent in and relevant to the Global South. To be fair, Dobson raises the issue of collective action (albeit in terms of voluntary activism) and stresses the asymmetry of the relationship (and the disparate ecological footprints) between the Global North and the Global South. Yet, somewhat paradoxically, given Dobson's explicit concern with justice, his concept of ecological citizenship leads to 'a focus on already powerful actors as the key protagonists' (Latta 2007b: 385), neglecting (at least in terms of active and responsible citizenship) the majority of the Global South. In other words, Dobson's concept of ecological citizenship does not appear to reach beyond the lifestyle politics of the Global North.

Finally, there are three significant themes that have been taken up by different authors, and which can also be taken as general critiques of his work. They are: (1) its androcentric (male-centric) bias or lack of sufficient attention to gender matters; (2) its anthropocentric (human-centric) nature; and (3) its potential for authoritarian politics (especially if the private sphere is opened to public scrutiny) that can result from ecological citizenship. These critiques, some of which apply also to mainstream conceptions of environmental citizenship, have often developed in the form of alternative or emerging theories of environmental citizenship. We will explore these perspectives in the final section of the chapter. In any case, and despite all the critiques, some of which he has directly responded to (Dobson 2012), it is important to reiterate that Dobson's work and, in particular, his conception of ecological citizenship continue to be 'the main point of reference in the debate on environmental citizenship' (Neuteleers 2010: 504).

Alternative and emerging perspectives

In this section we will outline several emerging and alternative theories and reflections on the greening of citizenship. These perspectives address some of the oversights of the theories explored thus far and propose interesting and radical ideas. Some of these inform actions, policies and initiatives that will be explored in the remaining chapters and have the potential to shape the future of citizenship in new and interesting ways. All of them contribute in their own way to the increasingly complex and rich field of environmental citizenship. The four approaches covered are: (1) the feminist approach; (2) the ecocentric approach; (3) the democratic approach; and (4) the neoliberal approach.

Gendering ecological citizenship: ecofeminist citizenship

Feminist theorists of citizenship have had little to say about the environment until recently. This is somewhat surprising given that feminist theorists have made a significant contribution to green political thought in the past few decades. Indeed, feminist ecological thinking, known as ecofeminism, has been and still is one of the most prominent schools of green political thought (Gaard 2011). Ecofeminism 'subverts the assumptions of other approaches by identifying the androcentric foundations at work in debates on the environment' (Smith and Pangsapa 2008: 42). That is, they reveal the male-centric perspective from which most ecological thought operates. Ecofeminist authors draw our attention to the close relationship between the domination of nature and the domination of women. They note how the consistent and long-standing association of men with the realm of the mind, reason and humanity, and women with the realm of the body, emotion and nature (Figure 3.4), has served over the centuries to justify the domination of men, simultaneously and complementarily, over women and nature (Plumwood 1992).

Ecofeminists have reacted against this characterization in different ways, but their work has often attached a considerable significance to the ethics of care and to women's experience of caring. This has translated into an ethics of care for the planet, famously theorized as 'earthcare' by Carolyn Merchant (1996). However, early notions of earthcare have failed to engage with citizenship, not least because 'ecofeminists are inclined to dismiss citizenship as the inherently exclusionary product of elite male minds' (MacGregor 2006b: 101). There is some truth to this

Figure 3.4 *'Mother Earth'*

Source: Flickr. Author: Digitum Dei.

claim, but that is gradually changing, as we have seen in Chapter 2. In line with this change, feminist articulations of environmental citizenship have begun to emerge. One of the most interesting contributions is the work of Teena Gabrielson and Katelyn Parady. Their work on 'corporeal citizenship' explores green citizenship as embodied practice, emphasizing 'the dynamic connectivity and co-constitutive interactions between human bodies and the nonhuman natural world' (Gabrielson and Parady 2010: 383). However, the most significant of the feminist articulations of ecological citizenship is the work of Sherilyn MacGregor.

MacGregor problematizes the strong association between motherhood and caring for the environment that is often made regarding women's environmental activism, and one that is still embraced by prominent ecofeminist activists and theorists (e.g. Shiva 2005). Instead, she argues that celebrations of earthcare as women's unique contribution to sustainability often neglect the importance of politics and citizenship in women's lives. She encourages ecofeminists to move 'from care to citizenship' (MacGregor 2004a), proposing a project of feminist ecological citizenship that affirms the practice of citizenship as an intrinsically valuable activity, albeit one that needs to be approached cautiously given the gendered division of labour that still operates in most societies. MacGregor also directs her analysis at the field of environmental citizenship, to argue that 'the emerging green discourse of "environmental citizenship" suffers from its lack of attention to gender relations' (2006b: 101). In short, her work brings citizenship into ecofeminism, and feminism into ecological citizenship. This double move comes to fruition and is particularly well illustrated in her monograph, *Beyond Mothering Earth* (2006a).

The main insight derived from MacGregor's feminist ecological citizenship comes arguably from the feminist insights into the politics of time and the gendered nature of the public/private divide. Her work brings to light the failure of environmental citizenship discourses to note the impact that the inevitable increases in time and effort created by green lifestyle practices will have on citizenly pursuits. Citizens who are involved in recycling, reusing, self-providing and making green ethical consumer choices on the home front will necessarily expend more time and energy on a personal level, in the private sphere. MacGregor notes that, since most of the domestic services are still performed by women, the more environmental citizenship is pushed into the private sphere, the more its burden will fall, disproportionately, on women. Moreover, in the absence of a general democratization of domestic life, the net effect of adding green duties onto (already time-poor) women will be to further restrict their time to participate in public life. In addition, green citizens are also expected to get involved in their community and participate in collective decision-making in a range of different political settings and institutions. In the present context, she argues, the twin emphasis on green lifestyle changes in the private sphere and greater participation in the green public sphere effectively results in asking women to do more (duty) with less (time). Last but not least, MacGregor points out that the social unsustainability of such demands on half of the population will ultimately undermine environmental sustainability (2006b). Her conclusion is as clear as it is emphatic: 'No (Ecological) Sustainability without (Gender) Justice'.

Naturalizing ecological citizenship: the natural contract

Theories of green citizenship are largely anthropocentric, even if this is typically a weak anthropocentrism. Their fundamental concern is the sustainability of the planet for the continuation of human life. However, some theorists argue for a more ecocentric approach that would entail the inclusion of animals and nature into the concept of citizenship. Some of these theorists work within the liberal and republican traditions (e.g. Curry 2000; Hailwood 2005). Others have taken the proposition of including the nonhuman world onto more radical territory. This is the case, for example, with Anna Bullen and Mark Whitehead, whose conception of 'sustainable citizenship' relies heavily on the conception of ecological citizenship elaborated by Dobson, but goes a step further. There are also radical ecocentric approaches that, whilst not typically associated with the field of environmental citizenship, have clear and direct implications for our conceptions of environmental citizenship.

Bullen and Whitehead draw on Dobson's work but differ from this in their inclusion of the nonhuman world. Their argument is that failing to include nonhumans assumes 'that the citizen predates the various bio-ecological processes which enable citizenship to even be conceived of in the first place' (Bullen and Whitehead 2005: 507). The implication is that we are natural beings (i.e. beings constituted by bio-ecological processes) before we can assert our condition of human beings, let alone assert our citizenship, that is, our condition of political beings. In the absence of such bio-ecological processes we cease to be. Since the natural world is a constitutive element of the human world, we must find ways to integrate both worlds into our conceptions of citizenship. The extension of ecological citizenship 'beyond the human' is also advocated by Matthew Hall, whose work espouses an ecological anarchism that 'identifies non-humans as fellow anarchists and activists, intelligently collaborating in the eco-anarchist defense of the Earth' (2011: 374).

The most radical and influential proposition is arguably the concept of a 'natural contract' formulated by Michel Serrès in *The Natural Contract* (1995). His point of departure is that nature is a political agent, an interlocutor in political life. He writes: 'Those who share power today have forgotten nature, which could be said to be taking its revenge but which, more to the point, is reminding us of its existence' (Serrès 1995: 29). That reminder is treated effectively as a political statement, nature's call for a natural contract. Similarly to the social contract, the powerful abstraction that serves to structure human societies, the natural contract would serve to structure relations between humans and nature. This would be 'a contract in which our relationship to things would no longer involve mastery and possession, but an admiring stewardship, reciprocity, contemplation, and respect' (Serrès 1995: 11). The contract would involve reciprocal roles, rights, responsibilities and relationships between nature as sustenance of humanity and humanity as stewards of nature.

Similar radical ideas can be found in the work of Bruno Latour, most notably in *We Have Never Been Modern* (1993) and *The Politics of Nature* (2004). Latour argues that nature is always already integral to the evolution of the dynamic assemblage we call society. He flatly rejects the distinction between humans and nature, which he associates with the philosophy of modernity, and this leads him to conclude that 'we have never been modern' (Latour 1993). Instead, he argues, we have always been social-cultural-natural hybrids. Modernity, rather than clarifying the picture, has contributed to the proliferation of such hybrid beings. In any case, this means that a politics focused on human interests alone is bound to fail, because our interests and those of nature cannot be truly separated. Here, the natural world is not conceptualized as a passive resource but as a participant in a 'dance of agency' that includes all natural and artificial things. Latour advocates the representation of all beings (human, nonhuman and artificial) in a 'Parliament of Things' (1993). A similar proposition, albeit coming from an ecofeminist perspective and with a strong pluralist undertone, is the 'Council of All Beings' proposed by Catriona Sandilands (2000).

These theories represent a radical take on citizenship, an institution that has always been about structuring political relations between humans.

Thus far, such theories have not gained much traction in the real world, but the notion of the 'rights of nature' has already informed some specific initiatives and policies (as we shall see later). This 'naturalizing' of ecological citizenship – which produces forms of ecocentric citizenship – brings up some important questions. The most important might be this: who can legitimately speak for nature or translate what nature is saying or trying to tell us, according to these theories? The answer, necessarily, must be humans. But, which humans? Thus far, the dominant answer has been experts on nature, such as biologists and other scientists. This would imply a significant shift of political power from traditional political authorities to scientific and technical experts – a shift that would force us to rethink not just the meaning of citizenship, but the meaning and operation of democracy. There is, of course, the possibility that other groups (e.g. indigenous peoples are sometimes mentioned in this context) could emerge as legitimate 'translators', and even the whole of society could perform that role through mechanisms of deliberate democracy. In any case, the naturalizing of citizenship, ecological or otherwise, is something we should pay attention to in the years ahead, not least because of the profound impact it could have on (democratic) politics and citizenship.

Democratizing ecological citizenship: the Global South

The potential threat to democratic politics posed by notions of ecological citizenship has been taken up by several authors from inside the field of environmental citizenship. Their focus is less on the radical ideas discussed in the previous section and more on the general direction of the field, including influential articulations such as the model of ecological citizenship proposed by Dobson. The main preoccupation is the very definition of ecological (and environmental) citizenship as 'citizenship for sustainability'. This instrumental understanding of citizenship is problematized most notably in the work of Alex Latta, often in collaboration with other authors, such as Nick Garside and Hannah Wittman.

Latta and Garside argue that the incorporation of the environment in the sphere of citizenship should not come at the expense of democracy. They write: 'The concept of citizenship contains a strong disruptive and democratic impulse that is arguably sidelined when it is conceived primarily as a vehicle for sustainability' (Latta and Garside 2005: 3). In view of this, Latta and Garside propose that 'the notion of ecological citizenship should not only inspire efforts to introduce ideas of sustainability into citizenship, but should also involve a critical application of citizenship practices and democratic ethics into sustainability' (2005: 3). In other words, the influence should cut both ways, rather than turn citizenship into a mere tool for environmental sustainability. These authors remind us of the potential for undemocratic impositions or green authoritarianism that can result from the opening up of the private sphere to public scrutiny to monitor environmental behaviour.

Latta also argues that scholars of ecological citizenship who share a concern for cultivating democratic tendencies in green politics should direct far greater attention to the actual spaces in which ecological citizens are being (and becoming) political on a daily basis. This has the advantage of infusing theoretical articulations of ecological citizenship with the pluralism and cultural diversity that has been largely, if not completely, absent. The other advantage is bringing into the picture citizens previously marginalized or treated as spectators, as is the case with the populations of the Global South. In addition, this approach enables the formulation of multicultural ideas of ecological citizenship that are sensitive to conceptions of human–nature relations amongst indigenous peoples. The need to ground theories of ecological citizenship on existing political spaces and practices is shared by an increasing number of authors, many of whom find examples of such democratic and pluralistic practices in Latin America (Latta and Wittman 2012).

The incorporation of indigenous cultures, values and peoples into theories of ecological citizenship brings with it its own challenges. There is a widespread but problematic perception that indigenous peoples live in harmony with nature. Whilst there is some truth to that claim, in the sense that their general impact on nature has been historically much less damaging than that of industrial societies, it does not automatically follow that all indigenous peoples are green or will embrace the environmental agenda embedded in notions of ecological citizenship. They might side, for a variety of reasons, with the developmental agenda rather than the environmental one – although sustainable development suggests that the two agendas are mutually compatible. In any case, placing democracy and pluralism at the heart of ecological citizenship enables the recognition of groups often ignored, including indigenous peoples, but also, more generally, the populations of the Global South.

Consuming ecological citizenship: sustainable consumption

The final theoretical reflections relate to the increasing influence of neoliberal thought on articulations of environmental citizenship, including ecological citizenship. To be sure, that influence is more pronounced in its practical manifestations (as we shall see throughout the text) than in the theoretical field. There are, however, some important implications for the theorizing of green citizenship derived from the increasing significance of market values and consumer culture in contemporary societies. The neoliberal ascendancy has translated into concepts and articulations of environmental citizenship that incorporate the environment into the notions of consumer citizenship and corporate citizenship. The most significant of these are the concepts of sustainable consumption and corporate environmental citizenship. However, to begin with, we must reflect briefly on another concept, one that is central to neoliberal thought and that impacts directly and profoundly on all notions of neoliberal environmental citizenship: property.

Neoliberal theorists argue that the best way to address issues of sustainability is by turning nature into private property and deploying markets as the solution to environmental problems. The privatization of natural resources makes individual owners responsible for the value of their property and thus guarantees their adequate care (Anderson and Leal 2001). This requires the pricing not only of natural resources but also of ecological services, that is, the services that nature provides to society in general and to the economy in particular (e.g. bee pollination of crops, forest absorption of carbon and wetlands purification of water). The treatment of nature as 'natural capital' and its integration into the dynamics of the market economy are best exemplified by developments such as carbon trading schemes and intellectual property rights regimes (Swanson 1995). The practical implications of this approach for environmental citizenship – located, first and foremost, within the concept of corporate environmental citizenship – are given ample consideration in Chapter 6. Here, it suffices to add that this model of citizenship is driven primarily by business corporations, albeit often with the acquiescence of international institutions such as the World Trade Organization (WTO) and the United Nations Environment Programme (UNEP).

In addition, neoliberal theorists argue that the most effective way in which citizens can embrace environmental citizenship is through their power as consumers, that is, embracing sustainable consumption. In essence, neoliberal environmental citizenship recasts the individual as a green consumer (rather than a green citizen). The critics of this approach focus on the negative impacts of neoliberal reforms on the environment, but they also highlight the distributional implications of the various forms of 'accumulation by dispossession' that follow the transformation of nature into private property (Bakker 2007). The most famous critique of the consumer approach to the environment is found in the work of Mark Sagoff. His position is neatly summed up in these words: 'We are citizens, not just consumers. Our environment requires citizen preferences, not just consumer preferences. As citizens, we need to protect nature, not just buy, sell, and consume it. It has a dignity, not just a price' (Hargrove 2001).

The extent to which sustainable consumption is more or less (if at all) compatible with notions of ecological citizenship will be an important matter of debate for years to come. The question has been explored by Gill Seyfang (2005). Seyfang suggests that sustainable consumption can promote ecological citizenship, but only if citizens are supported by institutional and political changes that create the appropriate social context. This would require governments to shift social values, set an example by developing green infrastructure, and force corporations to change their behaviour rather than rely on self-regulation. Her attempt to place consumption under the constraints of ecological citizenship is compelling, but relies on curbing the current neoliberal ascendancy. In the present economic and political context, the opposite seems to be happening. Citizenship is being consumed by market values, and active citizenship is often synonymous with shopping. In this sense, sustainable consumption might best be described as citizenshop. In any case, whether sustainable consumption will enhance sustainability or not will probably depend less on its compatibility with ecological citizenship and more on whether neoliberal ecological economics can actually deliver what it promises – a prosperous, yet greener future.

Summary points

- There is a strong perception that our current environmental predicament demands the 'greening' of our conception of citizenship. Many have addressed this challenge without abandoning conventional theories of citizenship, whilst others have pushed the concept significantly, sometimes radically.
- The mainstream theories of citizenship (liberal and republican) have incorporated the environment in a range of different ways, but mostly

through their traditional emphasis on rights (liberal approach) and duties (republican approach).

- The main point of reference in the theoretical field of environmental citizenship is the work of Andrew Dobson, in particular his conception of 'ecological citizenship'. This is a justice-based account of how we should live, based upon private and public action to reduce the environmental impact of our everyday lives on others.
- Dobson articulates a radical redefinition of citizenship that contrasts with traditional models in four key areas: it is non-reciprocal; it involves the private sphere; it emphasizes 'feminine virtues' (e.g. care); and it is non-territorial (i.e. includes future generations).
- There is a range of alternative and emerging theories that pay particular attention to different aspects of environmental citizenship often neglected by mainstream approaches.
- The most significant alternative theories show that gender matters, promote the inclusion of nature into citizenship and argue the need to place democracy and pluralism at the heart of all conceptions of environmental citizenship.
- Neoliberal ideas are also becoming influential in the field of environmental citizenship – although their influence thus far is felt much more strongly in the realm of practice.

Selected readings

Barry, J. (2006) 'Resistance is Fertile: From Environmental to Sustainability Citizenship', in A. Dobson and D. Bell (eds.), *Environmental Citizenship*. Cambridge, MA: The MIT Press, 21–48.

Bell, D. (2005) 'Liberal Environmental Citizenship'. *Environmental Politics*, 14(2): 179–194.

Dobson, A. (2003) *Citizenship and the Environment*. Oxford: Oxford University Press.

Gabrielson, T. (2008) 'Green Citizenship: A Review and Critique'. *Citizenship Studies*, 12(4): 429–446.

Jelin, E. (2000) 'Towards a Global Environmental Citizenship?' *Citizenship Studies*, 4(1): 47–63.

MacGregor, S. (2006a) *Beyond Mothering Earth: Ecological Citizenship and the Politics of Care*. Vancouver: University of British Columbia Press.

MacGregor, S. (2006b) 'No Sustainability without Justice: A Feminist Critique of Environmental Citizenship', in A. Dobson and D. Bell (eds.), *Environmental Citizenship*. Cambridge, MA: The MIT Press, 101–126.

Martinsson, J. and L. J. Lundqvist (2010) 'Ecological Citizenship: Coming Out "Clean" Without Turning "Green"'. *Environmental Politics*, 19(4): 518–537.

Seyfang, G. (2005) 'Shopping for Sustainability: Can Sustainable Consumption Promote Ecological Citizenship?' *Environmental Politics*, 14(2): 290–306.

Shiva, V. (2005) *Earth Democracy: Justice, Sustainability, and Peace*. London: Zed Books.

Online resources

Environmental Citizenship, The Goodenough Primer (2005). Available at: www.publicspace.ac.uk/environmentalcitizenship/PDF/ecprimer.pdf

Mark Sagoff, The Twin Failures of Ecological and Environmental Economics (uploaded 28 September 2011). Duration 13.31 mins. Available at: www.youtube.com/watch?v=71hrN4ce5Qw#t=13

Marti Kheel, 'Feminism and Animal Liberation: Making the Links' (2010). Duration 77 mins. Talk given at the Cease Animal Torture 2010 Conference held at California State University, Long Beach. Available at: http://vimeo.com/15088578. Marti Kheel is a writer in the fields of ecofeminism, animal advocacy and environmental ethics: http://martikheel.com/

Teena Gabrielson, Environmental Citizenship (uploaded 16 December 2010). Duration 7.51 mins. Available at: www.youtube.com/watch?v=58VwRvnKKb0

Vandana Shiva, Earth Citizenship Required Today (uploaded 10 September 2010). Duration 9.47 mins. Available at: www.youtube.com/watch?v=zyhUhviQScA

Vandana Shiva, Giving up Growth (uploaded 23 August 2010). Duration 3.55 mins. Available at: www.youtube.com/watch?v=2UlJvG7-okM&feature=related

Student activity

What kind of green citizen are you? Johann Martinsson and Lennart Lundqvist (2010) test the level and significance of citizen compliance with the model of ecological citizenship proposed by Andrew Dobson. They pose the question: does it matter what values we hold so long as our behaviour is green? Their study concludes that environmental attitudes are not essential to generate green behaviour. Do you agree with their conclusion? Before you read the article, write down a list of what you see as your environmental attitudes and values (i.e. do you consider yourself a green citizen?), and then answer the questions listed in their article's Appendix (p. 537). Do you get the sense that your attitudes and practices match up or do you sense a significant mismatch?

Martinsson and Lundqvist identify four main types of ecological citizen based on the two sets of answers: (1) Believers (citizens whose ecological practices are consistent with their positive attitudes towards the environment: green and clean); (2) Coverts (citizens whose ecological practice is inconsistent with their negative attitudes towards the environment: not green but clean); (3) Hypocrites (citizens whose ecological practice is inconsistent with their positive attitudes towards the environment: green but not clean); and (4) Diehards (citizens whose ecological practice is consistent with their negative attitudes towards the environment: neither green nor clean). What kind of citizen are you? Having done the test, do you think that practice, not attitudes and values, is what really matters? If so, what are the implications of such a conclusion for the field of environmental citizenship?

Discussion questions

- Which theoretical approach to environmental citizenship do you find more compelling and/or best reflects your position on the matter?
- Do you think the government should introduce some form of compulsory 'sustainability service'? If so, why? And what kind of services do you have in mind? If no, why not?
- What do you make of the 'natural contract'? Do you think nature can and should be incorporated as a political agent in our conceptions of citizenship?
- Do you think sustainable consumption in general, and sustainable shopping in particular, can promote ecological citizenship (see Seyfang 2005)?

Part II
Actions and practices

4 Environmental citizenship in action

> At first I thought I was fighting to save rubber trees, then I thought I was fighting to save the Amazon rainforest. Now I realize I am fighting for humanity.
>
> Chico Mendes

Citizenship cannot exist without citizens, particularly active ones. This chapter explores the relation between conceptions of citizenship and the actions of environmental citizens, in particular, those of environmental activists. The chapter cannot do justice to the thousands of campaigns and organizations that, in a more or less explicit fashion, bring together citizenship and the environment. Instead, the chapter provides a map to make sense of the myriad ways in which environmental activism is shaping citizenship by greening existing conceptions of citizenship and by inspiring new articulations of green citizenship. The focus will be on a selection of prominent movements, with attention paid to the different material and political contexts in which these operate, particularly those that differentiate between environmentalism in the Global North and the Global South. To that end, the chapter begins with a brief account of the framework that will be used here to classify environmental activism: 'the three posts'.

Contexts and frames: the three posts

The list of issues and concerns taken up by environmental activists over the years is long and varied, including amongst the most significant: industrial pollution, soil erosion, deforestation, nature conservation and climate change. Most issues, like industrial pollution and climate change, are common to the Global North and the Global South, while others, like deforestation and wildlife poaching, are more prominent in the Global South. However, what differentiates environmental politics across the world is not so much the list of issues but the different material and political contexts in which these issues manifest themselves. Indeed, once again: 'context is all' (Alfonsi et al. 2014). In material terms, the main contrast is the prevalent poverty of the Global South versus the prevalent affluence of the Global North. In political terms, the contrasts derive from the rule of law and representative governments that prevail in the consolidated democracies of the Global North versus the corruption and repression more typical of the fragile democracies and authoritarian regimes of the Global South. The concept of fragile democracy is used here to denote the lack of a fully-fledged democracy, with robust processes, mechanisms and structures that enable citizens to participate freely and openly in political activities, exercising their rights and duties in a political environment free from violence, intimidation, cronyism and/ or widespread corruption.

The different contexts translate into different dominant themes and priorities, and help to explain the different tenor of environmental activism in the Global North and the Global South. Affluence and democracy translate into a greater preoccupation with lifestyle politics, consumer rights and sustainability in the Global North. Poverty and repression translate into a greater preoccupation with livelihood politics, production rights and development in the Global South. In addition, the 'environmentalism of survival' of the Global South is much more radical than the 'environmentalism of affluence' of the Global North (Faber 2005). The expression 'environmentalism of survival' is typically used to denote the close association between environmental activism and the provision of basic needs and natural resources (upon which the satisfaction of those needs often depends) that characterize the 'environmentalism of the poor' (Martínez-Alier 2002). However, the expression is also used here to denote the political context of violence and intimidation that, literally, makes environmental activism a matter of life and death in many countries of the Global South.

These different contexts are particularly well captured in the typology of environmentalism known as 'the three posts' (Doherty and Doyle 2006): (1) postmaterial (typical of the Global North); (2) postindustrial (predominant in the Global North, but also present in the Global South); and (3) postcolonial (typical of the Global South). This typology will be used here to help us map the greening of citizenship by environmental movements. In recognition of the fact that environmental activism, as a form of political activism, is an expression of citizenship, there will be some references to postindustrial citizenship, postmaterial citizenship and postcolonial citizenship in this chapter – but these should be taken only as orientative categories, not recognized categories of citizenship. In addition, it is worth remembering that categories are conceptual tools and cannot neatly capture the complexity of reality. The three posts, for example, struggle to accommodate early expressions of environmentalism, particularly the conservation movement, which predates the postmaterial turn of the twentieth century. But, on the whole, 'the three posts' remain the most useful scheme to navigate the complex landscape of environmental movements, particularly since their explosion onto the scene in the 1960s.

Postindustrial environmentalism. This term refers to environmental activism associated with the impact of industrialization on the environment. Postindustrial environmentalism developed in response to the increasing level of waste and pollution in the urban centres, but also as a reaction against the nuclear industry, most notably in Western Europe and the Asia-Pacific. This is an anthropocentric movement that emphasizes the protection of human health through the struggle for a clean and healthy environment. The main manifestations of postindustrial activism are the anti-pollution, the anti-toxic waste and the anti-nuclear movements. This type of activism is commonplace in the Global North but has gradually extended across the world, following the increasing industrialization and urbanization of the Global South. Postindustrial environmentalism includes also the environmental justice movement, which has gained a global dimension with its condemnation of the double standards in the treatment of waste and pollution between the Global North and the Global South.

Postmaterial environmentalism. This term typically refers to environmental activism associated with the emergence of postmaterial values in affluent societies. The label is associated with the 'culture shift' in advanced industrial societies theorized by Ronald Inglehart (1977) and the 'hierarchy of needs' theorized by Abraham Maslow (1943). Postmaterial environmentalism is premised on the notion that once we have met our basic needs (e.g. food and shelter), we gradually become concerned with higher-level needs, or move from the concern with needs to the interest in wants. In environmental terms, this translates primarily into support for the conservation of nature as something to value in and of itself, but also for the purposes of leisure and aesthetic pleasure. The term postmaterial is also used here to encompass the nonmaterial (e.g. spiritual and aesthetic) expressions of environmental activism that predate the affluence-related postmaterial turn of the twentieth century. Postmaterial environmentalism is often associated with ecocentric positions (e.g. rights of nature) and developed in the Global North. The main and most influential manifestation of this type of activism is the nature conservation movement, known also as the wilderness movement. This movement originated in the United States and has subsequently been exported to rest of the world by conservationist non-governmental organizations (NGOs), such as Conservation International (CI), Nature Conservancy (NC) and World Wide Fund for Nature (WWF). The power and influence of these large international NGOs often translates into the imposition of the environmental priorities of the Global North on the Global South.

Postcolonial environmentalism. This term refers to environmental activism associated with the struggle against the economic colonization of the Global South by the Global North, but also to environmental movements that are part of the resistance against material poverty and political oppression in the Global South. Postcolonial environmentalism encompasses a wide range of movements, including the anti-dams, the anti-mining and the anti-deforestation movements. These movements bring traditionally marginalized sectors of the population to the fore of environmental politics, in particular peasants, women and indigenous peoples. Postcolonial movements integrate environmental issues into a broader political agenda that includes civil, political and social rights, indigenous rights, the rights of nature, human development and women's empowerment. In other words, the essence of this activism is political resistance, both against the political regimes and conditions of the Global South and against the domination of the Global South by the Global North.

Postindustrial activism and citizenship: North and South

The most explicit and direct connection between environmental activism and citizenship comes from postindustrial environmentalism, in particular from the anti-toxic waste movement and its focus on human health. Not surprisingly, the main impact on citizenship derived from postindustrial activism has been extending the protection of a particular set of human rights, namely, health rights. This protection is typically articulated in terms of the right to a healthy environment. However, the impact of postindustrial activism on citizenship goes well beyond the consolidation of health rights. In this section we will explore that broader set of impacts, with particular reference to what still is widely regarded as the most prominent and influential early expression of environmental citizenship derived from the modern environmental movement, the grassroots campaign that came to be known as Love Canal (see Box 4.1).

Box 4.1

Love Canal: a brief history

Love Canal refers to a New York neighbourhood built on an abandoned industrial canal that had been used by a subsidiary of Occidental Petroleum to dump over 20,000 tons of industrial waste between 1942 and 1953. Unaware of the origins of the site, the residents of this working-class area began to complain of odours and 'substances' appearing in their yards in the 1950s, but the area continued to be developed until 1978, when the pollution story broke in a series of articles published in the *Niagara Gazette*.

The news galvanized the neighbours, including the woman who would become the face of the movement, Lois Gibbs. Concerned with the health of her son and her neighbours, in August 1978 she co-founded the Love Canal Homeowners Association (LCHA), which struggled for more than two years for the relocation of the families of Love Canal. The campaign was ultimately successful, with President Jimmy Carter eventually ordering the evacuation of 900 families from Love Canal in October 1980. Love Canal led to the passage of the Comprehensive Environmental Response, Compensation, and Liability Act of 1980, better known as the Superfund. The resulting law created a federal programme paid into by waste generators to clean up hazardous waste sites throughout the United States, and Gibbs became known as the 'mother of the Superfund'.

Lois Gibbs went on to found the Citizens Clearinghouse for Hazardous Waste (CCHW) in 1981. She renamed it the Centre for Health, Environment and Justice, in 1997, in recognition of the fact that the anti-toxics movement had by then joined forces with the environmental justice movement, made famous by the publication of Robert Bullard's *Dumping in Dixie* (1990). Over the years, the CCHW has assisted thousands of grassroots groups with organizing, technical and general information, across the United States.

Resources

Centre for Health, Environment and Justice. Available at: http://chej.org/
The New York Times, Retro Report, 'The Love Canal Disaster: Toxic Waste in the Neighborhood' (published on 26 November 2013). Duration 11 mins. Available at: www.youtube.com/watch?v=Kjobz14i8kM

Postindustrial activism as liberal citizenship: the struggle for rights

Postindustrial activism tends to operate in ways that can be framed within classical conceptions of citizenship. In general terms, all forms of environmental activism, insofar as they illustrate the notion of citizenship as action, can be read as expressions of republican citizenship. However, this interpretation is problematic because it equates, always and by definition, liberal citizenship with passive citizenship. The reality of citizenship, even more so than the theory, is not that simple. Liberal citizenship can also manifest itself in the form of political activism, and hence as active citizenship. The key to determining the type of citizenship at work in postindustrial activism cannot be activism itself but the focus, purpose and values underpinning the actions of each particular group, campaign and movement. In essence, the focus on duties, the community and the public good is more suggestive of republican citizenship; whereas the focus on rights, the individual and private interests is more suggestive of liberal citizenship. In any case, we must keep in mind that the distinctions will always be a matter of emphasis, since most groups, campaigns and movements incorporate a wide range of elements that can be associated with various types of citizenship.

With those caveats in mind, it is not too difficult to see that the centrality of rights in postindustrial activism both reflects and reinforces the dominant modern articulation of citizenship, namely, liberal citizenship. The alignment with liberal citizenship can be seen first and foremost in the focus on health rights (i.e. on the impact of industrial pollution on human health) that is at the heart of postindustrial activism. This is clearly visible in the case of Love Canal, and well illustrated by the words of Lois Gibbs:

> At the heart of this movement is a concern for health combined with the desire for justice and human rights. People believe that no one has the right to make their families sick or their environment unsafe for any reason.
>
> (2011: 8)

The particular focus on the health of children and prospective mothers also illustrates the concern with the rights of future generations that has become central to articulations of environmental citizenship. The movement also campaigned for the right-to-know and was successful in getting the government to pass the Emergency Planning and Community Right-to-Know Act in 1986. This law gives citizens the right to know what chemicals are being stored, transported and disposed of at facilities and plants in their community. Finally, the alignment between postindustrial activism and liberal citizenship is also noticeable in the appeal to property rights that often accompanies campaigns against the installation of polluting industries and waste disposal facilities, and the claims for financial compensation that invariably follow incidents of pollution or contamination, as was the case in Love Canal (Fletcher 2002). Here, the link between citizen activism and property rights was evident in the very name of the main organization behind the campaign: the Love Canal Homeowners Association.

Postindustrial activism as glocal citizenship: the planet as backyard

Postindustrial activism has contributed to establishing the local and the global as important political spaces for the articulation of citizenship. In addition, the intersection between these two spaces has produced a new political space: the glocal. In this space, citizens see themselves and operate, simultaneously, as members of localized polities and global communities. The transnational networks that result from the coming together of local groups from around the world are expressions of glocal citizenship. The proliferation and increasing significance of these networks challenge the modern, automatic and exclusive association of citizenship with the national space, and point towards the local, the global and the glocal as additional political spaces increasingly relevant for the exercise of citizenship.

The emergence of the local as the main political space of environmental citizenship is typically associated with the anti-toxic waste movement, a movement informed by the notion that 'all politics are local' (Gibbs 2002: 105). Because of its focus on the local, the movement against toxic waste in general, and Love Canal in particular, came to be seen as expressions of not-in-my-back-yard (NIMBY) activism. This approach was criticized as being narrow and selfish, and ultimately ineffectual, resulting in local tactical victories at the expense of larger structural reforms and the type of national and global regulations required to dealing with toxic wastes in a systematic and predictable fashion. However, in time, local concerns and campaigns came to be identified with global issues and struggles, situated in everybody's backyard (Rootes 1999). Since then, the anti-toxic waste movement has expanded to include an increasing number of local communities, forming national and global networks, effectively transforming this movement into an expression of glocal environmental citizenship.

The shift to the global came mainly through the anti-nuclear movement and the realization that nuclear accidents cannot be limited in space and time. The notion that, in nuclear terms, the planet is everyone's backyard, including the backyard of future generations, has influenced the emergence of not-in-anyone's-back-yard (NIABY) activism. The focus on the global reflects and enhances the notion of cosmopolitan citizenship in general and cosmopolitan environmental citizenship in particular, both of which view the planet as the relevant political community to address environmental issues. This cosmopolitan and future-oriented vision of citizenship informs the environmental activism of some of the most renowned and influential environmental organizations, most notably Greenpeace, and is also central to green political parties, including the most successful and influential of them, the German Greens (Die Grünen). This vision is illustrated in the Charter of the Global Greens, the international network of green parties and political movements, which opens with the following statement: 'We, as citizens of the planet and members of the Global Greens'. Similar sentiments can be found in the 3rd Annual Green

Figure 4.1 *Anti-nuclear demonstration, Stuttgart (2011)*

Source: Flickr. Photographer: BÜNDNIS 90.

Oration, delivered on 23 March 2012, by the former leader of the Australian Greens, Bob Brown, whose opening address commenced with the words 'Fellow Earthians'.

Postindustrial activism as pluralist citizenship: the rights of 'minorities'

Postindustrial activism is not usually associated with pluralist notions of citizenship, mainly because its focus has been on public health and human rights. However, there are significant intersections between the two, particularly through the environmental justice movement. This movement brings marginalized groups to the forefront of citizenship struggles, in particular by highlighting how the distribution of environmental rights and risks reflects (and reinforces) existing prejudices and social inequalities, both in the Global North and between the Global North and the Global South. Postindustrial activists tend to deploy notions of egalitarian citizenship, thus situating the movement closer to liberal citizenship than the concept of differentiated citizenship typically associated with pluralist citizenship. However, both approaches illustrate the struggle for the rights of 'minorities' (often, in fact, the majority e.g. women) and their inclusion into substantive citizenship that drives pluralist conceptions of citizenship.

The most significant and widespread intersection between postindustrial activism and pluralist citizenship relates to matters of substantive citizenship and derives from the environmental justice movement. This movement originated in the United States in the 1980s as a response to the realization that polluting factories and waste sites were disproportionately located in poor and black neighbourhoods. The movement crystallized into a set of principles contained in the Declaration of Environmental Justice, the outcome document of the First Conference on Environmental Justice, held in 1982 in Washington, and was popularized following the publication of Robert Bullard's *Dumping in Dixie* (1990). This book was the first to document environmental racism, a phenomenon that has produced some of the most notable toxic locations in the United States, including the so-called 'Cancer Highway' in Louisiana. The struggle against environmental racism has become a subset of the environmental justice movement and has extended to other countries, most notably South Africa (Barnett and Scott 2007). The environmental justice movement has also gained a global dimension that translates into campaigns for global environmental standards and for putting a stop to the continuing use of the Global South as the dumping ground for the toxic waste of the Global North.

The adoption of the environmental justice discourse in the South has extended beyond the urban terrain and the postindustrial themes of pollution and environmental services typical of the movement in the North. In the South, the campaigns are simultaneously struggles for social justice, natural resources and land rights – placing the movement in closer alignment with postcolonial environmentalism. The contrast derives in part from the different communities involved – urban/suburban in the North, and rural, peasant and indigenous in the South. Here, environmental justice includes also the struggle to revive and revalorize indigenous ecological knowledge and facilitate the inclusion of indigenous peoples and perspectives in environmental decision-making (Rasch 2012). This last point positions the environmental justice movement closer to the pluralist notions of differentiated citizenship, in particular, indigenous citizenship.

The intersection between postindustrial activism and pluralist citizenship is also manifested in the significant involvement of women in the environmental justice movement at the grassroots level, including in positions of leadership. Yet, interestingly, that involvement has not translated into the incorporation of gender into the discourse of environmental justice, which continues to be dominated by references to colour and class (Buckingham and Kulcur 2009). Moreover, their involvement is often framed in the language of caring and motherhood – a language that is sometimes politically effective, but reinforces traditional gender stereotypes (Peeples and DeLuca 2006). In recent times, activists have begun to develop intersectional approaches to the practice (and teaching) of environmental justice (Di Chiro 2006). But for our purposes here, the campaign of Love Canal remains one of the most interesting illustrations of the complex gendered dynamics of postindustrial environmental citizenship.

Love Canal was a campaign led by women who used their roles as wives and mothers to articulate their demands for state action in resolving the environmental crisis in their local community (Hay 2009). The extent to which their maternalist discourse was a mere strategy to attract media attention or the product of essentialist notions of womanhood is disputed (e.g. Blum 2008). What is clear is that these women did not articulate a feminist conception of citizenship that openly challenged patriarchal structures and gender inequalities. Yet, their activism effectively undermined traditional family roles and exposed the potential costs to private life that can derive from women's participation in public life in patriarchal contexts. The time women spent as activists often led to family tensions and strained relationships, some of which ended in marriage break-ups, including that of Gibbs herself (Gibbs 2011). The personal price paid by these women and their families illustrates the 'politics of time' that underpins the exercise of citizenship, and vindicates the feminist calls to democratize the private sphere if one wishes to facilitate the participation of women in public life and their just integration into the realm of environmental citizenship.

Postindustrial activism as surrogate citizenship

Postindustrial environmentalism is typically associated with the Global North, but there is a rich and vibrant tradition of postindustrial activism in the cities of the Global South and the 'neither North nor South' region of Eastern Europe. In fact, environmental activism in general and postindustrial activism in particular have been instrumental in the development of citizenship outside of the Global North. Here, environmental activism tends to operate as a form of surrogate politics, particularly in the context of authoritarian regimes. This dynamic is facilitated by the fact these regimes operate with a narrow view of politics that considers green issues to be soft issues, and thus tend to tolerate green organizations. This somewhat apolitical view of environmental activism has enabled green organizations to become surrogate platforms for wider political reforms, including significant improvements in civil, social and political rights. This was the case in Eastern Europe during the Soviet era (Yanitsky 2012), and continues to be the case in the Global South.

Environmental activism in general and postindustrial environmentalism in particular have played a significant role in the struggle for democracy in much of the Global South. The case of Brazil is particularly interesting. Brazilian environmental activists were in the vanguard of the political movement that culminated in the return of the country to democracy in 1988 (Hochstetler and Keck 2007). Not only that, but their activism translated into the inclusion in the Brazilian Constitution of the most expansive environmental protections anywhere in the world at that time. Postindustrial activism continues to be a useful surrogate for traditional politics in countries like Iran, Burma and China. In Iran, the environmental movement – on the surface a typical postindustrial movement focused on pollution – is one of the main platforms of the reformist pro-democracy movement. The movement operates with a 'depoliticized politics' to avoid attention from the regime's hardliners while 'encouraging people to be active citizens in the social and political sphere' (Fadaee 2011: 85). The most significant ongoing illustration of postindustrial activism as surrogate citizenship is currently taking place in China. Here, environmental activism has becomes a 'safe' form of exercising and deepening citizenship, since environmental causes are not seen to pose a direct political threat to the Chinese Communist Party. In the past few years, citizens have managed to pressure political authorities to stop industrial projects on environmental grounds, placing these activists in the vanguard of political change in China (Mao 2014).

Postmaterial activism and citizenship: North on South

Postmaterial environmentalism is not typically associated with citizenship. Indeed, its fundamental concern with nature and animals seems at odds with matters of citizenship. However, there are significant intersections between the two, mainly derived from the language of rights, e.g. rights of nature and animal rights. This language invokes notions of liberal citizenship, but the extension of rights to animals and nature suggests a radical departure from classical liberal citizenship. Indeed, postmaterial activism, through its most prominent manifestation, the nature conservation movement, poses some of the most formidable challenges to the traditional notion of citizenship. In this section we will explore the impact of postmaterial activism on citizenship in a range of different areas, with particular attention to the ways in which this activism operates as the imposition of the environmental values of the Global North on the Global South.

Postmaterial activism and substantive citizenship

The most obvious way in which postmaterial environmentalism impacts on citizenship is through the creation of national parks – the foremost and most long-standing achievement of the nature conservation movement. Their very name, national parks, is a clear indication of their explicit link with the dominant conception of political community in modern times: the nation. National parks have always been part and parcel of the creation of national identities, and as such they reinforce the conception of national citizenship. The relation between national parks and national identity extends across the world, but is particularly powerful in the country that pioneered their creation, the United States. Since their early days, national parks became integral to the cultural life of the American nation. Their significance as a source of spiritual renewal and national pride continues to these days, most recently on display in the award-winning six-part documentary entitled *The National Parks: America's Best Idea* (2009).

The designation of certain areas as national parks almost invariably has implications in terms of substantive citizenship. National parks have often been created at the expense of local and indigenous peoples, whose displacement from their homelands confirms and reinforces their status as second-class citizens in most countries. These dynamics can be traced back to the very origins of national parks in the United States, where the establishment of Yosemite Park in 1864 entailed the eviction of native inhabitants from their homelands (Spence 1996). The promotion of national parks has remained central to environmental activism in the United States and has been pursued with particular fervour by the largest American-based NGOs, most notably the so-called big three: Conservation International (CI), Nature Conservancy (NC) and the World Wide Fund for Nature (WWF). The influence of these and other

Figure 4.2 *Yosemite National Park*

Source: Wikimedia Commons. Photographer: Chensiyuan.

large international NGOs has translated into the imposition of their nature conservation agenda onto the Global South.

The extension of the conservation agenda around the world has created thousands of 'conservation refugees', particularly in African and Asian countries, including Tanzania, Uganda, Botswana, Kenya, South Africa, India and Thailand (Dowie 2009). This term refers to the millions of farmers, herders and hunters evicted or displaced from lands and forests to make way for parks, sanctuaries and wildlife reserves. In recent times, conservationist organizations have made efforts to introduce participatory forms of conservation designed to improve the efficacy of conservation measures, while ensuring they do not become a form of environmentally infused neo-colonialism (Agrawal 2005). However, entrenched power structures continue to limit the extent and impact of such initiatives (Paulson *et al.* 2012).

The globalization of the conservation agenda has led to the emergence of two new expressions of cosmopolitan environmental citizenship: conservation volunteering and ecological tourism. Conservation volunteering is a relatively new practice that enables citizens to volunteer in a range of wildlife and conservation projects. The practice offers environmental citizens the opportunity to make a difference and to contribute to conservation projects that otherwise might not be funded. In the global context, these schemes enable the performance of 'cosmopolitan global environmental citizenship' (Lorimer 2010). The extent to which this practice will help to enhance the status of the citizens of the Global North at the expense of the local peoples of the Global South is difficult to establish at this point, but the potential is certainly there. The picture is a lot clearer regarding the practice of ecological tourism, or ecotourism. In general terms, the promotion of global ecotourism has enhanced the green (and global) citizen credentials of the northern tourists who can afford to visit the national parks and protected areas of the Global South. In this context, the displaced locals become second-class citizens in relation to their compatriots, and third-class citizens in relation to the affluent tourists from the Global North. In short, globalized postmaterial activism creates spaces that can engender a cosmopolitan environmental feeling amongst the citizens of the Global North, but this often comes at the expense of the substantive citizenship of the local peoples of the Global South.

'Parked animals': postmaterial activism and the rights of nature

The other major impact on citizenship derived from postmaterial environmental activism comes from the campaigns for the protection of animals, especially when such protection is framed in terms of animal rights and/or the rights of nature. The best-known illustration of the animal rights discourse associated with the conservation movement is the anti-whaling movement. This movement dates back to the 1970s and has produced some of the most visible and long-standing environmental campaigns of all times, and spawned some notorious NGOs, most notably the Sea Shepherd Conservation Society. However, the most significant dynamics regarding matters of citizenship originate from the protection of land-based animals, in particular, large mammals (e.g. gorillas, lions, tigers, elephants, rhinos and bears). Indeed, as is the case with citizens, when it comes to nature conservation, some animals are more equal than others. Clearly, size matters. There is something about large fauna that makes them attractive to humans, particularly to the citizens of the Global North.

The protection of land-based animals is articulated through the creation of national parks and other protected areas, and as such is part of the dynamics we explored in the previous section. Ramachandra Guha and Joan Martínez-Alier (1997) provide a particularly illustrative example of how the setting aside of wilderness areas to protect a particular species, the Bengal tiger, has impacted on the indigenous communities and the poor peasant populations of India. Their study refers to the network of parks known as Project Tiger. The authors note how the designation of tiger reserves entailed the physical displacement of existing villages and their inhabitants, with their management requiring the continuing exclusion of peasants and livestock. In essence, the project puts the interests of the tigers ahead of the interests of the peasants living in and around the reserves – effectively treating the latter as second-class citizens vis-à-vis the tigers that come under the protection of Project Tiger (Figure 4.3).

The extent to which the clash between animal rights and human rights can manifest itself is best illustrated by the 'shoot to kill' policies against poachers in Kenya's wildlife parks that led to the death of over 100 people between 1989 and 1991. The policy was endorsed at the time by the WWF and the International Union for Conservation of Nature (IUCN), amongst others (Doherty 2002: 214). Their support for such policies can be better understood if we consider that the survival of

Figure 4.3 'Citizen Tiger'?

Source: Getty Images. Photographer: Tratong.

protected animals is directly linked to their own reason for being as
organizations. This makes conservation organizations act as players with
their own survival interests rather than partners interested in the survival
of local and indigenous peoples (Doyle and McEachern 2008: 142).

The protection of animals in the nature conservation movement is
typically framed in terms of the 'rights *of* nature'. However, this frame
hides the political dynamics behind the establishment of such spaces
and the employment and enjoyment particular humans derive from the
existence of 'parked animals'. In fact, the protection of animals in parks,
national or otherwise, is best understood in terms of the 'right *to* nature',
namely, the human right to enjoy nature and biodiversity. In this sense,
the 'rights of nature' discourse poses no significant challenge to the
traditional conception of citizenship, and can be seen as an additional
human right that can be easily accommodated into liberal citizenship, or
liberal environmental citizenship. In keeping with this understanding,
and given the global dimension of these dynamics, the 'rights of nature'

discourse effectively asserts the 'right to nature' of the Global North at the expense of the local and indigenous peoples of the Global South.

Simian citizenship: The Great Ape Project

The discourse of animal rights opens up the space for a much more radical idea: the treatment of animals as citizens. The most serious challenge of this kind is the Great Ape Project, established in 1994, and spearheaded by Peter Singer and Paola Cavalieri. The Great Ape Project is an international organization of experts, including ethicists, primatologists and anthropologists, including Jane Goodall, Richard Dawkins and Jared Diamond, who advocate extending basic legal rights to nonhuman great apes (henceforth: great apes), namely: chimpanzees, bonobos, gorillas and orangutans. They propose three civil rights: the right to life, the protection of individual liberty and the prohibition of torture. These rights have been included in a Declaration of the Rights of Great Apes, whose advocates wish it to be adopted by the United Nations. The declaration is based on the notion that humans and great apes form a 'community of equals' because we all share the same fundamental attributes: we are intelligent animals, with a rich social, emotional and cognitive life. In light of this, its advocates argue that denying great apes basic civil rights because they belong to a different species is a form of discrimination, which they call speciesism, akin to other forms of discrimination that have been used to exclude some humans from citizenship, most notably blacks (racism) and women (sexism).

The Great Ape Project advocates the establishment of a representative body of experts to act as guardians of the great apes, similar to existing trusts that operate under the United Nations. This body would be entrusted with the defence and management of 'the first nonhuman independent territories' and would also play a role in 'the regulation of mixed human and nonhuman territories' (Cavalieri and Singer 1993b: 311). The establishment of territories where nonhuman great apes can live freely necessitates rethinking the concepts of sovereignty and citizenship, amongst others. The most serious attempt thus far to grapple with the political implications of this project has been Robert Goodin, Carole Pateman and Roy Pateman's essay entitled 'Simian Sovereignty' (1997). The authors identify two ways of implementing 'simian sovereignty', both requiring the creation of internationally protected reserves or autonomous territories for the great apes. The first would consist in 'debt for nature' swaps, that is, the cancellation of foreign debt to a country in exchange for the provision of certain protections for the great apes already living in that country. The second would be a conventional 'purchase' of land, which would become protectorates or 'mandated territories', administered by an international governmental organization, such as the UNEP or an NGO, such as the WWF. The trustees would safeguard the interests and rights of the great apes as they would those of humans incapable of acting for themselves, e.g. children and the intellectually disabled. Their role would be to guarantee the survival and autonomy of the great apes without further human interference (Goodin *et al.* 1997: 839).

The extension of legal rights to great apes signifies the extension of legal citizenship to a group of nonhumans, albeit with a limited number of rights at the moment, i.e. basic civil rights. However, the question arises: why stop at basic civil rights? In a community of equals, there is no good reason not to complement those rights with basic social rights, e.g. health rights. The subsequent question is: why stop with the great apes? Once the human/nonhuman species barrier is crossed, what prevents the logical extension of civil and social rights to other nonhuman complex animals? Indeed, the authors of the declaration present this extension of rights to great apes as a cautious step, given their relatively small number, but recognize that some 'would like to see a much larger extension of the moral community, so that it includes a wider range of nonhuman animals' (Cavalieri and Singer 1993a: 1). In any case, the process has already begun. In 2008, Spain became the first country to commit to the Great Ape Project, granting rights to great apes on the grounds that humans share approximately 99 per cent of their active genetic material with nonhuman primates (MacGregor 2011: 276).

The extension of legal rights to great apes and other nonhuman animals opens up the potential for conflict between humans and nonhumans when interests collide. But the political conflict will not derive from the relationship between humans and nonhumans, but from the relationship between different groups of humans – in particular, between those designated to represent nonhuman animals and those whose interests and rights come into conflict with the nonhuman animals represented by such bodies (or with the interests of those in such bodies). These schemes place significant – even sovereign – authority in the hands of particular groups of experts, most notably scientists and biologists. In theory, this makes perfect sense: who better than scientists and biologists to understand and guarantee the well-being of nonhuman animals and nature in general? However, given the contentious history of nature conservation, many are afraid of the implications that such schemes could entail, especially for the local and indigenous peoples of the Global South.

The words of US conservation biologist Daniel Janzen play into such fears by providing a chilling insight into the imperialist drive that could be facilitated by such initiatives: 'If biologists want a tropics in which to biologize, they are going to have to buy it with care, energy, effort, strategy, tactics, time and cash' (cited in Guha 1997a: 14). These words illustrate what some see as the inherent danger with the 'rights of nature' discourse; that it can be used to undermine democratic politics in favour of a techno-scientific elite with authority to control potentially large chunks of territory, mostly (albeit not exclusively) in the Global South. To the extent that these initiatives might involve the participation of green activists and conservation volunteers, they could further the association of cosmopolitan (and simian) environmental citizenship with the imposition of the values and interests of the Global North on the Global South.

Postcolonial activism and citizenship: South vs North

The impact of postcolonial activism on citizenship derives, first and foremost, from the centrality of political resistance that characterizes environmental activism in the Global South. This makes postcolonial movements and campaigns expressions of insurgent citizenship, defined as: a form of active participation in social movements whose purpose is 'the *defense* of existing democratic principles and rights' and/or 'the *claiming* of new rights that, if enacted, would lead to an *expansion of the spaces of democracy*' (Friedman 2002: 77, original emphasis). In addition, postcolonial activism brings to the fore two groups who historically have been excluded from citizenship and continue to be marginalized or treated as second-class citizens in many countries, namely, women and indigenous peoples. In this sense, postcolonial activism is closely associated with pluralist citizenship. Finally, in general terms, postcolonial movements are better at integrating humans and nature, because they reflect the different cosmologies and the closer relation between humans and the natural environment found in the Global South. Postcolonial articulations of environmental citizenship tend to be more comprehensive than their postindustrial and postmaterial counterparts, which tend to focus on humans (in the case of postindustrial citizenship) or nature (in the case of postmaterial citizenship). In this section, we will explore the main intersections between postcolonial activism and environmental citizenship, with specific attention to a novel conception of citizenship inspired by the rubber tappers movement of Brazil.

Postcolonial activism as indigenous environmental citizenship

Environmental activism in the Global South is often associated with the activism of indigenous peoples, and their resistance to mega-projects and industrial practices of resource extraction that threaten their homelands and livelihoods. The most prominent and widespread movements with large indigenous participation are the anti-mining, the anti-dams and the anti-deforestation movements. The participation of indigenous peoples in these movements has been crucial to their gradual inclusion into citizenship in their respective countries as well as to the struggle to develop a set of global environmental human rights. In addition, some of the campaigns with a strong indigenous presence have challenged the radical separation between humans and nature and the anthropocentrism at the heart of classical conceptions of citizenship.

The struggle for indigenous rights within a classical citizenship framework is well illustrated by the campaign of the Ogoni people against the environmental degradation of their land caused by oil extraction in Nigeria. This campaign combines calls for the right to a clean and healthy environment with demands to be included in the distribution of natural resources. This and similar campaigns bring together the language of indigenous rights and environmental rights, under the overarching discourse of human rights, contributing towards the formulation of environmental human rights (Hancock 2003). The struggle to obtain full citizenship extends also to demands for the recognition of their rights as indigenous peoples (i.e. indigenous rights). The two approaches can operate side by side but they refer to different conceptions of citizenship. The first is a struggle framed in the language of inclusion that can and has been easily incorporated into the liberal narrative of universal human rights and the egalitarian articulations of pluralist citizenship. The second is a struggle framed in the language of recognition that centres on difference and invokes notions of cultural and indigenous citizenship, that is, one that illustrates pluralist notions of differentiated citizenship.

The recognition of indigenous citizenship has been aided in no small part by the alliance between environmental activists and indigenous peoples. The alliance has been successful in several countries, most notably in

Brazil. Here, the political alliance between the two groups was crucial in achieving the inclusion of indigenous rights and environmental rights in the 1988 Constitution. These constitutional rights have been central to the struggle for Brazilian environmental causes, particularly in the region of the Amazon, most notably in the campaigns against the building of large dams, such as the Belo Monte dam (Cao 2014). The campaign has galvanized national and international action in support of indigenous rights, including popular figures such as James Cameron and Sigourney Weaver. Cameron even produced a short film, entitled 'A Message from Pandora' (2010), which compares the battle to stop the Belo Monte dam with the struggle of the Na'vi people against the destruction of nature in *Avatar* (2009). The alliance between indigenous and environmental activism is also prominent in India, particularly in the struggle against the construction of the Narmada dam, which counts amongst its most popular supporters the author Arundhati Roy.

The relationship between indigenous activism and environmental citizenship is well illustrated in a recent study by Raoul Rasch of the local resistance against mega-projects in the region of Huchuetanango, Guatemala. The study shows the twofold way in which indigenous peoples employ citizenship in the struggle against the exploitation of their environment. On the one hand, they call for inclusion in the existing model of citizenship by demanding their basic rights, like any other citizen, to participate in the decision-making process. On the other hand, their activism creates new categories of inclusion and formulations of citizenship enabled by their cultural identity. Thus, for example, Mayan lawyers 'have argued that according to Mayan spirituality, the extraction of minerals in open-pit mines is equal to wounding Mother Earth' (Rasch 2012: 178). They argue that, in contrast to the individualist and economic approaches to natural resources of non-indigenous peoples, indigenous peoples consider their relationship with 'Nature' and 'Mother Earth' part of their cultural identity. This allows them to claim the right to their territory by demanding respect for their cultural rights as indigenous peoples. This articulation of their claims goes beyond classic egalitarian notions of citizenship, reflecting instead a culturalist as well as a relatively ecocentric notion of citizenship. Here, the politics of nature and citizenship become intertwined in ways that differ greatly from conventional notions of environmental citizenship in the Global North. The coming together of indigenous difference and environmental activism produces what might be called insurgent indigenous environmental citizenship.

Postcolonial activism as ecofeminist citizenship

Women have been present in a wide range of environmental movements and campaigns around the world, but have played a particularly significant role in the postcolonial movements of the Global South. Their presence has been most notable in the forest movement, where they have often occupied positions of leadership and constituted a large proportion of the participants. The centrality of women in postcolonial environmentalism has inspired and reflected ecofeminist formulations of citizenship. Two of the most distinctive and influential illustrations of ecofeminist citizenship are the Chipko Andolan movement (from India) and the Green Belt Movement (from Kenya).

The movement that brought the women of the South to the front of global environmental politics was Chipko, an environmental movement that emerged in India in the 1970s. The Chipko movement was a reaction to the increasing deforestation that threatened the survival of forest communities in the region of the Himalayas. In 1977, a group of women adopted the strategy of 'hugging the trees' in order to save them from the loggers. Their tactic inspired the expression 'tree huggers' that is used often to mock environmentalists. Yet their actions reflected a strong stance in defence of the trees and their communities, which depended on the forest for their whole way of life (i.e. fuel, fodder, timber, medicines and food). The women deployed the language of motherhood in the articulation of their resistance, invoking along the way the notion of Mother Nature (Rangan 2000). This maternalist and naturalist discourse attracted accusations of essentialism, but their stance inspired a generation of women to engage in environmental politics and resist the incursion of industrial forestry into the habitats of indigenous peoples across the world (Moore 2011).

The other major influential ecofeminist movement is the Green Belt Movement (GBM), founded in Kenya in 1977 (see Box 4.2). The GBM has escaped the association with essentialism, largely because of the way it was conceptualized by its founder, Wangari Maathai (Figure 4.4). She placed the emphasis on women's empowerment, but operated with a complex understanding of the concept (Presbey 2013), and articulated a comprehensive vision and mission for the GBM (Maathai 2004a). Indeed, the GBM is much more than an ecofeminist movement. Maathai appreciated the interdependence of all aspects related to citizenship: rights and duties, civil rights and social rights, the individual and the collective, material and environmental conditions, and the political

Box 4.2

The Green Belt Movement: Wangari Maathai

The Green Belt Movement (GBM) is a grassroots, non-governmental organization based in Kenya that focuses on environmental conservation, community and capacity building. The GBM was founded in 1977 by Wangari Maathai, following the planting of seven trees on Earth Day to honour Kenyan women environmental leaders. The goal of the organization was to combat soil erosion and deforestation by creating public greenbelts and fuel wood plots by local people, especially women, in the spirit of self-reliance and empowerment (Muthuki 2006: 84). Planting trees was seen as a practical solution to myriad problems, as well as an expression of political resistance. Maathai viewed tree planting both as a good in and of itself but also as 'an entry point – a way in which women could discover they were not powerless in the face of autocratic husbands, village chiefs and a ruthless president' (Lappé and Lappé 2004). In Maathai's words, 'the tree became a symbol for the democratic struggle in Kenya'.

Maathai paid a heavy personal price for her activism. She received death threats, was tear-gassed, jailed and at one point clubbed unconscious by riot police. In addition, her advocacy for women's rights did not sit well with her husband who, in seeking divorce, complained that she was 'too educated, too strong, too successful, too stubborn and too hard to control' (Muthuki 2006: 85). Not surprisingly, they divorced. On the positive side, her work had a massive impact on the ground and received widespread recognition. Over her lifetime, the GBM planted over 50 million trees in Kenya. Importantly, the movement gradually extended beyond Kenya. In 1997, the GBM established a Pan-African Green Network to share their approach with other African countries. Maathai joined forces with the UNEP in 2006 to urge the planting of a billion trees worldwide – a target that was reached within a year. Her work was recognized with plenty of prizes, most notably the Right Livelihood Award (often called the Alternative Nobel) in 1984 and the Nobel Peace Prize in 2004. Since her death in 2011, Maathai has remained an inspirational figure. Her legacy includes a story she used frequently in her speeches: the story of a hummingbird. With this story, she encourages every single one of us, no matter how small and insignificant we might feel, to get involved and 'do the best we can'.

Resources

Wangari Maathai: 'I will be a hummingbird' (uploaded 11 May 2010). Duration 2 mins. Available at: www.greenbeltmovement.org/get-involved/be-a-hummingbird. Also available at: www.youtube.com/watch?v=IGMW6YWjMxw
The Green Belt Movement. Available at: www.greenbeltmovement.org/

Figure 4.4 *Wangari Maathai, founder of the Green Belt Movement*

Source: Taking Root. Photographer: Martin Rowe.

dimension of all these issues. The movement operates with the notion that these are interrelated issues that must be tackled as part of one complex and integrated front. In other words, struggles for social justice, women's rights and the environment are all tied together in her work and that of the GBM. Maathai was awarded the Nobel Peace Prize in 2004. The statement released by the Nobel Committee praised her work for 'sustainable development, democracy and peace', reflecting the multidimensional articulation of citizenship that often characterizes environmental activism in the Global South and that is palpable in the extraordinary work of Maathai and the GBM.

Florestania: forest citizenship in the Brazilian Amazon

Postcolonial activism has inspired the creation of a novel conception of citizenship: *florestania*. The word is a Portuguese neologism formed by the combination of two words: *floresta* (forest) and *cidadania* (citizenship). The term was coined by Antônio Alves Leitão Neto, an adviser to the state government of Acre, a region in the Brazilian

Amazon. However, this conception of citizenship came about as the result of the rubber tappers movement of Brazil that originated in the 1970s, as a reaction to the increasing deforestation of Acre. The movement was led by Chico Mendes, a formidable figure who brought national and international attention to the plight of the rubber tappers and the deforestation of the Amazon. His tragic death at the hands of a hired gun turned him into an 'eco-martyr' and the icon of the struggle to save the Brazilian Amazon.

The rubber tappers movement inspired a new model of development for the region, built on a new land-tenure regime: extractive reserves. This regime aligned biodiversity conservation and local forest people's land rights, enabling community livelihoods by promoting the sustainable extraction of forest products, rather than clearing the forest for agriculture. The first extractive reserves were established in the early 1990s, marking the beginning of a longer-term political trajectory spearheaded by the rubber tappers that led to the adoption of a model of 'forest-based development' at the municipal and state level. The concept of *florestania* developed by the regional government as part of this new regime of forest management was premised on a commitment to the principle of agricultural sustainability, but also to redressing the many facets of social exclusion suffered by forest dwellers. Indeed, the term *florestania* reverses the implied emphasis on *cidade* (city) within the concept of *cidadania* (citizenship), denoting the intention of extending citizenship to previously excluded forest residents (Schmink 2011: 142).

This articulation of citizenship reflects and intersects with a range of conceptions of citizenship, providing another excellent illustration of the complex relations between environmental activism and citizenship in the Global South. In keeping with the social liberal understanding of citizenship, *florestania* serves to extend citizenship rights to attenuate the social inequality inherent in capitalist market-oriented economic development policies. However, the concept came to mean more than the incorporation of forest dwellers into the existing political system. In addition to the extension of social rights (e.g. education, health care, housing and employment), *florestania* includes also 'the creation of new rights and cultural sensibilities that not only combat material deprivation but also combat the cultural rules that ignore the rural poor as subjects and as bearers of rights' (Schmink 2011: 142). This is illustrated, for example, in the commitment of the 'forest governments', as they came to be known, to support and respect the cultural knowledge and history of forest peoples, including the rubber tappers and indigenous communities.

In this way, *florestania* becomes an expression of insurgent citizenship derived from an ecosocial movement that emerged from the forests of the Amazon.

Yet, at the same time, *florestania* is also an illustration of neoliberal citizenship at work. The neoliberal dimension of the concept relates above all to the value and treatment of the forest as an economic resource. In line with this vision, the government of Acre transformed *florestania* into an ambitious development agenda set out to construct roads and bridges, support the private sector and subsidize controversial forest management proposals. But this treatment of the forest as an economic resource had always been present in the historical identity of the rubber tappers. Thus, in a sense, *florestania* simply reinforced and transformed that identity into 'a modernizing project oriented to national and global market opportunities' with the 'emphasis on individual rights and market mechanisms' (Schmink 2011: 153). In the final analysis, '*florestania* effectively served to manage social differences and to depoliticize social movement struggles by extending social rights . . . and by absorbing grassroots organizations and leaders as active participants in a project that reflects the commercialization of nature' (Schmink 2011: 153). In short, *florestania* became the perfect expression of ecological modernization and neoliberal environmental citizenship on the outskirts of the Brazilian Amazon.

Perhaps, the most revolutionary aspect of *florestania* is the conception of political community implicit in the term. The political centrality of the forest in this conception of citizenship effectively transforms forest residents into a political community of their own, based on their residence in and relationship with the forest, and with the corresponding rights and duties associated with that membership. Thus, despite the fact that the rubber tappers movement was largely a state-based affair, limited to the state of Acre, the concept of citizenship it inspired carries with it the potential to bring about a new political community within the Brazilian territory: the Amazon forest. The suggestion of the Amazon being a separate territory alarms many Brazilians, who are sensitive to the threat to national sovereignty implicit in any talk, especially at the international level, about the global significance and fragile state of the Amazon. Arguably, for the local, national and global citizens who support struggles to protect the Amazon forest, references to this territory are simply shorthand to refer to a region of the world they feel needs special protection – a region that just so happens to be mostly located in Brazil. In any case, and in view of the popularity of planting trees as an expression of environmental citizenship, perhaps there is something to be said about extending the concept of *florestania* to the whole planet – even if only symbolically – and begin thinking about Earth as a global forest in which we are all residents, with rights and duties regarding trees, that is, global forest citizens of planet Earth.

Past the posts: food citizenship and the politics of food

The final section of this chapter will be dedicated to the exploration of food citizenship and the politics of food. The significance of such politics cannot be underestimated. To a large extent, and as the saying goes: we are what we eat. Not only is food essential to human survival, and thus a constant reminder of our biological nature, but what and how we eat also say much about our cultural identity. Moreover, food plays a major role in our ecological footprint – a key pedagogical tool in the teaching of green citizenship that will be explored in Chapter 7. This section will revisit the intersection between environmental activism and citizenship through the three posts. This approach will illustrate the complex, diverse and multidimensional articulations of food citizenship, which both replicate and go beyond (i.e. cannot be neatly contained by) the three posts. In addition, the section explores the concept of agrarian citizenship, with particular attention to its most prominent manifestation, the grassroots peasant movement La Vía Campesina (LVC).

Environmental activism and food citizenship: past the posts

The politics of food manifests itself through a range of environmental movements, most notably the movement against genetically modified food (or anti-GM food movement), the slow food movement and the right-to-food movement (including its most prominent articulation, the food sovereignty movement). These movements are expressions of 'the three posts' but they also illustrate how the politics of food in general and conceptions of food citizenship in particular cut across or go 'past the posts'. This is particularly the case with the anti-GM food movement, which has postindustrial, postmaterial and postcolonial manifestations, and cuts across the North/South distinction. Nevertheless, even in this case, the distinction remains relevant. The anti-GM movement and the politics of food in general still take different forms and shapes around the world, with emphasis on the politics of survival or livelihood politics (focused on food production) in the Global South and on the politics of affluence or lifestyle politics (focused on food consumption) in the Global North.

Postindustrial activism and food citizenship. The best illustration of the intersection between postindustrial activism and food citizenship is the anti-GM food movement, particularly as it operates in the Global North. This movement is part of the wider movement against genetically-modified organisms (GMOs) that emerged as a reaction to the introduction of genetic engineering in the 1970s. In the North, the movement centres on the risks to human health derived from the consumption of GM foods. The concerns are mainly articulated in the language of health rights typical of postindustrial activism (Buttel 2005). This, as we have seen, invokes the notion of (social) liberal citizenship. However, the movement is also increasingly associated with calls for the adequate labelling of products containing GMOs. In fact, in affluent societies, the anti-GM food movement has largely become an expression of consumer-driven activism centred on demanding information for consumers ('the right to know'). This focus on consumer rights – when not consumer duties, i.e. 'the duty to know' – illustrates and contributes to the wider shift from liberal to neoliberal citizenship in the Global North (De Tavernier 2012). The anti-GM movement also takes on postmaterial expressions that resonate with the discourse of the rights of nature, particularly when concerned with the harmful effects that toxins can have on biodiversity. In addition, some of the expressions of this movement are associated with postmaterial values related to cultural practices, such as cooking and eating, that invoke the language of cultural rights associated with (multi)cultural citizenship.

Postmaterial activism and food citizenship. The best illustration of the intersection between postmaterial activism and food citizenship is the slow food movement, born in northern Italy in 1986. Slow Food International, the formal not-for-profit organization, numbers over 100,000 members around the world, and has grown into a global grassroots movement involving millions of people in over 160 countries. The slow food movement is a response to the homogenizing effects of the fast food culture promoted by neoliberal globalization (which some argue is nothing short of cultural imperialism) by 'promoting local foods, flavors, and cultures in a way that promotes social equality and environmental sustainability' (Merrett 2007: 1613). The movement is primarily a lifestyle movement, often focused on consumption, although it has important points of intersection with the concept of food sovereignty. The slow food movement promotes small-scale farming, biodiversity, sustainable agriculture, fresh food and a sense of community. This generates a form of 'food citizenship' that encourages participation and community-building, reflecting notions of republican and communitarian citizenship that go beyond the discourse of consumer rights typical of neoliberal citizenship. Their articulation of the local and the global makes this a 'glocal' movement, that is, a global effort to preserve local agriculture and culinary traditions. The focus on agricultural and culinary diversity also aligns this movement with notions of multicultural citizenship – or of what might be better termed multiculinary citizenship.

Postcolonial activism and food citizenship. The best illustration of the intersection between postcolonial activism and food citizenship is the right-to-food movement. This movement calls for the consideration of food as a fundamental human right – as already recognized in the Universal Declaration of Human Rights (Article 25). The right to food is central to food citizenship in the Global South, where malnutrition and hunger are still significant issues, particularly in South Asia and Sub-Saharan Africa. However, the movement is not exclusive to these regions, but is a global movement articulated by organizations such as La Vía Campesina (see Box 4.3). The right-to-food and food citizenship in general have gained prominence in recent times with the increasing purchase of land by foreign governments and corporations in countries of the Global South, most notably in Latin America and Africa (Golay and Biglino 2013). These purchases raise concerns about food sovereignty, an issue that promises to become central to food citizenship in the twenty-first century. The right-to-food and the anti-GM food movements are often in alignment politically – even though the two are presented as being in fundamental tension by those who support scientific solutions to global hunger. Indeed, the anti-GM food movement is one aspect of the struggle for food sovereignty in the Global South. Postcolonial activists view GM food (and GMOs in general) as an attempt to impose corporate models of development, food production and consumption across the world, reflecting the neo-colonial political dynamics of 'North on South'. The right-to-food movement and its most prominent articulation, the food sovereignty movement, are closely aligned with the notion of agrarian citizenship and the activism of LVC.

Agrarian citizenship: cultivating ecological citizenship

The concept of agrarian citizenship encompasses the political and material rights and practices of rural dwellers, and has been invoked as the solution to the so-called 'metabolic rift' between human societies and the natural environment (Wittman 2009a). This rift has intensified in recent times with the corporate takeover of agriculture by multinational corporations. This takeover has taken several forms, but of particular significance is the use of property rights (by means of patents) to gain control over seeds and thus extend corporate control over agricultural production. The patenting of seeds produces a profound shift in power from peasants to corporations. The corporate control of seeds poses the greatest threat to agrarian citizenship, potentially transforming agrarian citizens (e.g. peasants) into corporate subjects. In addition, these technologies result in a decrease in traditional varieties that can undermine biodiversity and multiagriculture. In this context, the act of saving seeds has become a political act – an expression of insurgent 'ecological citizenship' (Phillips 2005).

The concept of agrarian citizenship has been primarily associated with the activism of La Vía Campesina. LVC seeks to harmonize agriculture with the environment by challenging capitalist and industrial agricultural practices. They condemn and oppose the privatization of land, the current global land grabs, the treatment of food as a commodity, and the financial speculation with food 'futures'. They also reject the market-based approaches that promote 'green agriculture by partnering with leading agribusiness' and the claims that 'corporations can play a major role in supporting a transition to green agriculture' made by the UNEP, most recently in their report entitled 'Towards a Green Economy' (2011). In their view, the 'green economy' is nothing more than an exercise in 'greenwashing'. Instead, LVC demands profound structural changes to our economies, advocates the end to the commodification of nature and the defence of the commons, and calls for collective action against 'the state-corporate architecture', most notably against the giant multinational chemical and agricultural biotechnology corporation Monsanto.

La Vía Campesina operates with a model of citizenship that brings nature back into the politics of food. However, despite its evident ecological dimension, its focus is on human rights, and more specifically on the rights of peasants. These are made explicit in their signature document, the Declaration of the Rights of Peasants (2009). The declaration applies the term peasant to any person engaged in agriculture, cattle-raising,

Box 4.3

La Vía Campesina and food sovereignty

La Vía Campesina is an international grassroots peasant movement that comprises about 150 local and national organizations from 70 countries across Africa, Asia, Europe and the Americas. The movement brings together millions of peasants, small and medium-sized farmers, landless people, women farmers, indigenous peoples, migrants and agricultural workers from around the world. Together this represents about 200 million farmers. Its international committee rotates location every four years, and is currently based in Indonesia. The document that informs the work of La Vía Campesina is the Declaration of the Rights of Peasants (2009).

La Vía Campesina is an autonomous, pluralist and multicultural movement, independent from any political, economic or other type of affiliation. Their principal objective is to develop solidarity and unity amongst small farmer organizations in order to promote gender parity and social justice in economic relations, the preservation of land, water, seeds and other natural resources, sustainable agricultural production based on small and medium-sized producers, and food sovereignty. They strongly oppose corporate-driven agriculture, advocating instead small-scale sustainable agriculture as a way to promote social justice and human dignity.

The organizing principle for all their actions is the concept of food sovereignty, developed by members of La Vía Campesina and brought to global attention at the World Food Summit in Rome in 1996. They define food sovereignty as the right of peoples to healthy and culturally appropriate food produced through sustainable methods and their right to define their own food and agriculture systems. Food sovereignty prioritizes local food production and consumption, to ensure that the rights to use and manage lands, territories, water, seeds, livestock and biodiversity are in the hands of those who produce food and not of the corporate sector. In essence, food sovereignty asserts the principle that food is a basic human right and that participatory democracy is fundamental to its realization.

Resources

The Declaration of the Rights of Peasants (2009). Available at: http://viacampesina.net/downloads/PDF/EN-3.pdf

La Vía Campesina. Available at: www.viacampesina.org

pastoralism or handicrafts related to agriculture or a related occupation in a rural area. The document asserts the universality, indivisibility and interdependence of all human rights, but the focus of their activism is the 'right to food' or more precisely the 'right to feed oneself' (i.e. food sovereignty). Central to the realization of this right are two other rights: land rights and the right to seeds. However, LVC's engagement with these rights is not through a liberal human rights discourse but through the republican language of the common good. Their articulation of agrarian citizenship extends rights to peasants as part of a collective, based on a sense of community, not as individuals. Similarly, the land in general and seeds in particular are seen as a commons, not commodities. Seeds are the patrimony of humanity, part of the global commons, with peasants as their custodians. This is a model of citizenship premised on collective rights and the duty to exercise those rights in defence of the common good (Wittman 2009a).

In terms of political community, agrarian citizenship – and the activism of those who support this model of citizenship – refers to a range of political spaces: local, national, global and glocal. In theory, agrarian citizens should expect state protection for local rights 'but would also depend on local and global social networks and traditional ecological knowledge of agrarian conditions to enact those rights' (Wittman 2009a: 808). However, the central space of agrarian citizenship is the local. This is the space in which agrarian citizenship is ultimately lived and enacted – after all, agrarian citizenship is rooted in the land. LVC notes how corporations have no attachment to any given place, moving their production around the world through global outsourcing, without any consideration other than profit. They extract value and move on to the next source of value. Instead, peasants are rooted, which gives them reasons to nurture, to care for nature and the land. This focus on the local and the agency of peasants benefits both people and nature.

The emphasis on the local is complemented by the emphasis on the seasonal. Indeed, the two are closely interrelated. The protection of the local production requires the consumption of locally available produce, rather than imported foods, which sometimes come from the other side of the world. This emphasis on the seasonal has the advantage of addressing the metabolic rift between humans and nature, making citizens more aware of the natural seasons and rhythms of nature. In addition, the focus on the local and the seasonal can contribute significantly towards the fight against climate change, not least by reducing 'food miles', and thus the amount of energy that goes into transporting food around the world to supply out-of-season produce, particularly to the supermarkets of the Global North. This relates directly to the concept of the ecological footprint (or, in this context, the 'ecological foodprint') – a popular pedagogical tool for environmental citizenship that we will explore in Chapter 7.

The concept of agrarian citizenship can (and should) be extended to encompass urban agricultural initiatives. Indeed, urban agriculture is emerging as a significant expression of ecological citizenship, especially in the form of community gardens. There is an increasing number of studies suggesting that community gardens are becoming significant political spaces for growing engaged ecological citizens, especially in the Global North (e.g. Turner 2011). Community gardens and other urban agricultural initiatives, such as organic food networks and farmers markets (Figure 4.5), are helping to reconnect people with food, nature and community (Firth *et al.* 2011). These initiatives are expressions and incubators of ecological citizenship, providing opportunities for social learning about and public participation in food production and consumption practices and decisions (Travaline and Hunold 2010). The social learning that takes place in urban agricultural spaces has the potential to educate city residents in two ways. First, by learning

Figure 4.5 *Adelaide Farmers Market, South Australia, Australia*

Source: Private Collection. Photographer: Seán Mullarkey.

about their food (where, how and by whom it is grown) people may be able to make more informed decisions about their food system. Second, participating in food growing can teach people how to become effective citizens. Moreover, these initiatives can keep alive (protect and promote) alternatives to the dominant corporate food regime epitomized by agricultural corporations and large supermarkets. But we must be wary not to expect or ask too much of these initiatives. Their relatively small-scale and generally 'apolitical' (or 'lifestyle') nature means that urban agriculture can be (and often is) at the same time 'radical, reformist and garden-variety neoliberal' (McClintock 2014). In short, urban agricultural initiatives can be seen as expressions of agrarian citizenship that cultivate ecological citizenship in urban areas – but are not immune to the dominant neoliberal framework that shapes green 'lifestyle politics' in the Global North.

Summary points

- Different material and political contexts translate into different expressions of environmental activism and citizenship in the Global North and the Global South.
- Contrary to the notion that environmentalism is a matter for the affluent (a luxury only the rich can afford), some of the most revolutionary articulations of green citizenship are expressions of insurgent citizenship located in the Global South.
- The different forms of environmental activism (postindustrial, postmaterial and postcolonial) intersect with citizenship in a range of interesting ways, sometimes producing new articulations of environmental citizenship (e.g. agrarian citizenship).
- The main contribution of postindustrial activism to environmental citizenship is the emphasis on environmental human rights, particularly health rights. This type of activism originated in the Global North but has extended across the world, following the increasing industrialization and urbanization of the Global South.
- Postmaterial environmental activism poses some of the most significant challenges to traditional citizenship, mainly through the discourse of the rights of nature. This discourse originated in the Global North but has been exported to the rest of the world by powerful conservationist international NGOs, and often translates into the imposition of the environmental priorities of the Global North on the Global South.

- The main impact of postcolonial environmentalism on environmental citizenship derives from the participation of traditionally marginalized groups: women and indigenous peoples. The activism of these groups is often multidimensional but essentially defined by political resistance, both against the regimes and conditions of the Global South and against the domination of the Global South by the Global North.
- Environmental activism outside of the Global North often operates as a surrogate for politics and citizenship struggles. The resultant articulations of environmental citizenship tend to be much more comprehensive than those found in the Global North.
- The politics of food illustrates the complexity and multidimensional character of environmentalism and environmental citizenship, and food citizenship will shape much of what it means to be a (green) citizen in the twenty-first century.

Selected readings

Cavalieri, P. and P. Singer (1993) 'The Great Ape Project – and Beyond: Why the Project?' Available at: www.animal-rights-library.com/texts-m/cavalieri01.pdf

Gibbs, L. ([1982] 2011) *Love Canal and the Birth of the Environmental Health Movement*. Updated edition. Washington, DC: Island Press.

Lorimer, J. (2010) 'International Conservation "Volunteering" and the Geographies of Global Environmental Citizenship'. *Political Geography*, 29(6): 311–322.

Maathai, W. (2004) *The Green Belt Movement: Sharing the Approach and the Experience*. New York: Lantern Books.

Mendes, C. (1989) *Fight for the Forest: Chico Mendes in His Own Words*. London: Latin American Bureau.

Phillips, C. (2005) 'Cultivating Practices: Saving Seed as Green Citizenship?' *Environments Journal*, 33(3): 37–49.

Schmink, M. (2011) 'Forest Citizens: Changing Life Conditions and Social Identities in the Land of the Rubber Tappers'. *Latin American Research Review*, 46(4): 141–158.

Wittman, H. (2009a) 'Reworking the Metabolic Rift: La Vía Campesina, Agrarian Citizenship, and Food Sovereignty'. *The Journal of Peasant Studies*, 36(4): 805–826.

Online resources

A história de Chico Mendes [In Portuguese] (published on 22 August 2013). Duration 43 mins. Available at: www.youtube.com/watch?v=JoTHmdqz6lw

BBC Earth, Launching Project Tiger (published on 5 October 2011). Duration 2.32 mins. Available at: www.youtube.com/watch?v=nhSipS4d5og

Citizen Gardening in Public Spaces (published on 22 July 2013). Duration 7.55 mins. Available at: www.youtube.com/watch?v=2S09VrUPkHw

DAM / AGE (2003) Arundhati Roy's campaign against the Narmada dam project in India (uploaded 12 January 2011). Duration 50 mins. Available at: www.youtube.com/watch?v=QQ2iViE31bc

James Cameron, A Message from Pandora (uploaded 30 August 2010). Duration 3.11 mins. Short feature on the fight to stop the Belo Monte dam in the Brazilian Amazon. Available at: www.youtube.com/watch?v=RjfLyGTXSYo. The full length video is available here: http://vimeo.com/28181753 (duration 20 mins).

Should Apes Have Rights? (uploaded 14 February 2011). Duration 32.52 mins. Available at: www.youtube.com/watch?v=Pz6VYhGIj-Q

The Legacy of Chico Mendes (published on 4 November 2013). Duration 6.23 mins. Available at: www.youtube.com/watch?v=GzTddMd7Xz4

Student activity

Explore the citizenship profile of a green party or NGO. Select a green movement, NGO or political party you want to find out more about, and explore their website to identify the conception of environmental citizenship within which they operate. For example, you might explore their mission statement (which indicates basic values), video galleries and annual reports.

Explore the relation of that content with the basic elements of citizenship: membership, rights and duties (individual and collective). What kinds of rights are defended, and for whom? What kinds of duties are defended, and for whom?

Does the organization reflect moderate or radical conceptions of environmental citizenship? Is its perspective anthropocentric or ecocentric? Do you think the organization you have chosen reflects a liberal, republican, cosmopolitan, feminist, ecological, neoliberal conception of citizenship? Or a combination of them?

Discussion questions

- Which political space do you think should be prioritized when it comes to enacting green citizenship: the local, the national or the global?
- Should the United Nations approve the Declaration of Rights for Great Apes? If so, why? If not, why not? See www.projetogap.org.br/en/

- Do you think forests are political spaces? Or are they simply natural spaces? What would be the implications of thinking of forests as political spaces?
- Do you see food as a citizenship issue? If so, why? If not, why not? What are the main implications of conceptualizing food in terms of citizenship?

5 Governing environmental citizenship

> Government: 1. the group of people with the authority to govern a country or state; 2. the system by which a state or community is governed; 3. the action or manner of controlling or regulating a state, organization, or people.
>
> *Oxford English Dictionary*

Governments have become acutely aware of the need to address environmental concerns. Their specific policies and programmes vary significantly from country to country, but, across the world, governments are adopting the language of citizenship to frame their response to environmental issues. This chapter explores the main ways in which governments are deploying and shaping environmental citizenship as part of the greening of government that has taken place in recent decades. We will consider how this process has evolved within a shifting historical context, from the top-down 'government approach' of the early decades, to the network-based 'governance approach' that has become the dominant paradigm since the 1990s, and the 'governmentality approach' that is gaining momentum in the twenty-first century. These different forms of green(ing) government will be illustrated with examples from around the world, with particular attention to their articulation in the Global North and the Global South. We begin with an introduction to the three main forms of governing modern societies, including environmental matters, and the neoliberal context in which these forms ('the three govs') have evolved in recent times.

The three govs: government, governance and governmentality

Government, commonly understood, is a relatively simple concept. However, the practice of governing, particularly in late modernity, is a far more complex and subtle matter that requires engaging with two additional concepts: governance and governmentality. The three concepts are essential to understand the different ways in which political authorities articulate the relationship between citizenship and the environment. We will explore 'the three govs' separately but it is important to keep in mind that these forms of governing are not mutually exclusive.

Government: sovereignty and the rule of law

Simply put, government refers to the exercise of political authority over a community. In modern times, government is associated primarily with the nation-state and the concept of sovereignty (i.e. exclusive and supreme power over a given territory and population). In addition, modern government has also become synonymous with the rule of law – a form of ruling that relies on legislation to determine what is allowed and what is prohibited. Government thus understood reflects a top-down approach to governing, where political authorities exercise 'command and control' over the general population. This form of governing rests on the social contract between the state and the citizenry. In exchange for their loyalty, the state protects its citizens. Over time, the state has taken up more duties as part of that contract, from guaranteeing basic freedoms (civil rights) to enabling political participation (political rights) to providing social rights (welfare state).

The mechanisms and institutions underlying this form of governing are many and varied. In consolidated democracies, political power must be exercised with the formal backing of the citizens, which requires, at a minimum, regular free and fair elections and effective mechanisms of accountability. By contrast, in corrupt, dictatorial or authoritarian regimes, political authority does not derive so directly from popular authority and/or fails to comply with basic standards of accountability. In any case, all contemporary governments invoke the law and claim to act on behalf of their people, albeit with more or less credibility, in order to legitimize their rule and exercise their authority.

The term government continues to be the official and popular way to refer to the formal political authorities of a country, but in recent times a new term has emerged to describe the current (and increasingly hegemonic) operation of government: governance.

Governance: governing by stakeholders

The top-down approach to government has been challenged since the neoliberal turn of the late 1970s, coming under renewed pressure following the collapse of the Soviet Union. The demise of communism vindicated the reformist agendas of Margaret Thatcher in the United Kingdom and Ronald Reagan in the United States. Their campaign against 'big government', coupled with transnational economic globalization and developments in communications technology, facilitated the emergence of networks of private interests and corporations as significant political actors, particularly since the 1990s. Since then, it has become common to speak of governance rather than government as the best way to describe the complex and diverse range of actors, networks and regimes governing most aspects of contemporary societies, including the environment (e.g. Evans 2012).

This change in terminology signifies an important transformation in the way contemporary societies are governed, from an almost exclusive reliance on the sovereign state, to the participation of multiple actors in the process of governing, including technical experts, private companies and corporations, non-governmental organizations (NGOs), and civil society in general. These different actors come together through networks, and are often consolidated into regimes: 'institutions possessing norms, decision rules, and procedures which facilitate a convergence of expectations' (Krasner 1983: 2). The move towards governance has not undermined the significance of the law, but has made its formulation and implementation a much more cooperative approach. In short, this represents a shift from hierarchical and sovereignty-based governing to horizontal and network-based governing.

The main point of contention is the extent to which the shift from government to governance signifies a transfer of power from the state to non-state actors and, following from this, whether that transfer of power would constitute a positive or negative development. The optimists argue that this is indeed a positive development, one that enables a more pluralistic participation in policy-making, empowering citizens and effectively making the governing of societies more democratic (Pierre and Peters 2000). In contrast, the critics argue that this optimistic understanding of governance elides the structural power imbalances between the different sectors involved in the process and the technocratic dynamics that structure this form of governing. These critics interpret governance as the transformation of government into a depoliticized system of management, that is, into a purely technocratic activity dominated by unelected experts and managers who control the agenda, the policy-making process and its ultimate implementation (e.g. Sørensen 2002). The mechanics of this form of governing are so powerful, argue its critics, that they even trap political authorities, to the point that these abandon or subsume their broader political agendas and principles in favour of expert knowledge and managerial decisions. In short, governance replaces traditional political rule with managerial rule and the rule of experts.

Governance has crucial implications for the meaning and practice of citizenship, especially regarding the issue of citizen participation. The critical take on governance suggests that average citizens are not empowered by this form of governing, since their participation is invariably framed by the priorities and processes set up by relevant experts and managed by administrators. Thus, whilst the participation of citizens is real, and might even increase significantly, their actual capacity to have a substantial input and effect substantial change is minimal. In fact, and importantly, citizens gain responsibility without power, that is, by being part of the process, citizens become responsible for its outcomes, even though they had little real control over the process and thus over its ultimate outcomes. This critical understanding of governance is often associated with the analytical deployment of governmentality – a concept that is also reflective of a distinctive and increasingly influential form of governing.

Governmentality: self-regulation, or the conduct of conduct

Governmentality refers to a form of indirect and subtle ruling that shapes human behaviour in ways that mask the mechanics of power and thus achieves seemingly free and consensual compliance but without conscious and public deliberation. The term derives from the work of Michel Foucault, the French philosopher who popularized the idea that 'power is everywhere', diffused and embodied in discourse, knowledge and 'regimes of truth' (Rabinow 1991). Foucault used the term to shift the focus of political analysis from formal laws and institutions to the technologies of power that shape, direct and regulate our individual beliefs, desires, lifestyles and actions. He argued that coercion and regulation are not the only means used to produce ordered, predictable, manageable behaviour in modern societies (Foucault 1991). Instead, modern political power operates through more subtle and diffuse mechanisms designed to instil in each individual techniques of self-control (or self-regulation). In other words, governmentality represents a shift from the rule of law (a set of codes that is explicit and visible) to the rule of conduct (a set of codes that are implicit and shape will formation).

Governmentality, as a technique of government, has become increasingly associated with neoliberalism. Indeed, governmentality has become the preferred technique of neoliberal authorities. This makes perfect sense given the centrality of individual responsibility and self-regulation in neoliberal thought. Some authors have made the connection explicit, arguing that neoliberalism is 'a form of governmentality within which individuals discipline themselves to use their freedom to make responsible choices' (Bevir 2011: 466). In this form of governing, citizens are expected to 'examine and govern themselves so as to improve their lives in ways that benefit themselves, their community and the state' (Bevir 2011: 466). Political authorities still exist, but govern in a different way, *through* citizens rather than *over* citizens, shaping human conduct through the formation of will, ultimately inducing self-government.

There are significant synergies between governance and governmentality, particularly in the context of neoliberal societies. They are both forms of governing that operate through diffuse power and indirect ruling, and bring civil society into the operation of power. The best way to illustrate the difference between 'the three govs' might be with an example (Box 5.1).

Environmental government: green(ing) constitutional frameworks

The main approach to address environmental matters has been governmental regulation. This has resulted in thousands of pieces of legislation regulating economic activities that impact on the environment, protecting specific species and habitats, creating national parks, etc. The increasing volume and importance of environmental legislation in recent decades have been accompanied by the proliferation of environmental agencies, departments and ministries, both in the Global North and the Global South. Government legislation does not adopt the language of environmental citizenship explicitly, but contributes to its realization indirectly by providing an additional tool for the exercise of environmental activism. In addition to general legislation, more and more countries are introducing environmental provisions into

Box 5.1

Governmental approaches to forest management

Picture a government that wants to address the problem of unsustainable logging of forests, and assume it also wants to comply with international agreements to protect international tropical forests as well as its own old growth forests, which are being heavily logged to supply the paper industry and for furniture and other timber goods. The government can take three approaches to tackle the problem:

1. *The Government Approach*: The government prohibits, or heavily limits, the logging of its forests outright, and applies protective legal or fiscal mechanisms to encourage the development of quick growing and harvestable timber forests. Tariffs on foreign imports of tropical timber products could be applied. The production and import/export of timber products would need to be audited by governmental authorities and breaches of law enforced through normal legal channels.

2. *The Governance Approach*: The government initiates a negotiation process involving all stakeholders with an interest in forestry and forestry-related industries and products. Invited parties include non-governmental organizations (NGOs) such as wildlife and consumer organizations, labour unions, business organizations, forest owners, the timber and paper industries, and other interested businesses and industries, as well as relevant technical specialists such as forest product auditors, regulators, biologists, etc. The idea is to establish a voluntary code of practice in which a 'chain of custody' certification process is set up, whereby it can be demonstrated where all forestry products derive from at all steps of the production process. Producers and manufacturers who cannot comply with the minimum standards cannot receive the certification. The government sets up its own standard and, because forest management is an international issue, becomes involved in similar efforts at international level.

3. *The Governmentality Approach*: Environmental campaigns by NGOs and interest groups and awareness-raising campaigns run by the government and governmental authorities heighten sensitivity to the many ethical issues surrounding indiscriminate logging of old growth and tropical forests. Citizens are encouraged to care for the forests. The existence of the national forestry standard provides consumers with a sense of confidence that ethical standards are being met in the production of forest and paper products. Many timber and paper products are also prominently labelled as having received the tick of approval of the Forest Stewardship Council, the international not-for-profit organization that promotes sustainable forestry and sets standards for certification and labelling. These products also receive the prominent endorsement of the World Wide Fund for Nature (WWF). Both endorsements act as powerful marketing tools vis-à-vis products that do not have these endorsements. Consumers internalize that consumption of the appropriately labelled products is the right and proper thing to do, and may (or may not) act accordingly. Producers react to market signals. NGOs and governments monitor and report. The future of the forest is largely (if not ultimately) decided by market forces.

Resources

Forest Stewardship Council. Available at: https://ic.fsc.org/
The Montréal Process. Available at: http://montrealprocess.org/
US Forest Service. Available at: www.fs.fed.us/forestmanagement/

their constitutions. The inclusion of environmental rights and duties in constitutional texts signals an explicit engagement with the language of citizenship with the potential to shape the future of environmental citizenship in significant ways. In this section we will explore the implications derived from the constitutionalization of the environment.

General trends and implications: membership, rights and duties

The inclusion of environmental provisions in the supreme law of the land has significant advantages when it comes to environmental protection and environmental activism. For a start, this gives environmental provisions the highest rank amongst legal norms, trumping every statute, administrative rule and court decision. Addressing environmental concerns at the constitutional level thus means that environmental protections are more firmly rooted in the legal order and cannot be easily altered by a simple change of political majorities. In addition, the constitutionalization of the environment has a powerful symbolic and educational value, contributing to its elevation to the highest regard in societies, influencing and guiding public discourse and people's behaviour (Brandl and Bungert 1992).

The number of countries with environmental provisions in their constitutions has increased rapidly over the past 40 years. In the year 2000 more than 70 countries had constitutional environmental provisions of some kind, and in at least 30 cases these took the form of environmental rights (Hayward 2000). Since the trend began, no newly promulgated constitution has omitted references to environmental principles and many older ones are being amended to include them (Hayward 2005). The vast majority of constitutions contain relatively short provisions, mainly in the form of statements of public policy (i.e. programmatic statements that serve to guide or frame government actions when it comes to formulating environmental legislation). Some, however, include more substantive and detailed environmental provisions. The examination of these constitutional provisions reveals a significant range of articulations of environmental citizenship, but also some fundamental similarities, to which we now turn our attention.

Membership: present and future, but still national

There is no sense of cosmopolitan or global citizenship in environmental constitutional provisions. In other words, these provisions do not contemplate the spatial projection of citizenship beyond the territorially bounded political community of the nation-state. This is hardly surprising given that modern constitutions are the social contracts between nation-states and their citizens. There is, however, a temporal stretching of that political community, with significant implications for the meaning and practice of citizenship. This stretching takes place through the inclusion of future generations in constitutional environmental provisions. Their inclusion implies a duty towards unborn citizens that breaks with traditional conceptions of citizenship in two main ways. First, their inclusion effectively undermines the contractual nature of citizenship, given that future generations cannot agree to the contract (even though we can assume they would be happy that past generations endeavoured to leave behind a healthy environment). Second, their inclusion creates a duty for present generations that breaks the reciprocal nature of traditional citizenship, given that future citizens cannot reciprocate (although they can 'reciprocate' by endeavouring to leave a healthy environment for 'their' future generations).

Rights and duties: a matter of emphasis

The majority of constitutions balance republican and liberal traditions, with duties allocated both to the state and the citizenry. The variations are typically a matter of emphasis, largely explained by the political culture of the country and the historical context in which the texts were elaborated or amended. Thus, for example, the 1978 Spanish Constitution states that: 'Everyone has the right to enjoy an environment suitable for the development of the person as well as the duty to preserve it' (Article 45.1). In almost identical fashion, the 1982 Portuguese Constitution states: 'Everyone shall have the right to a healthy and ecologically balanced human environment and the duty to defend it' (Article 66.1). The Portuguese Constitution then goes on to list a series of measures that are 'the duty of the State' (Article 66.2). Similarly, the 1982 Turkish Constitution includes all three, in a single article: the citizen's rights, the citizen's duties, and the state's duties (Article 56).

The contrast between republican and liberal provisions is best illustrated by comparing the current constitutions of Greece and South Africa. The first is eminently republican, and statist; the second is liberal, of the social variety. The 1975 Greek Constitution makes no explicit reference to environmental rights and instead states that: 'The protection of the natural and cultural environment constitutes a duty of the State'

(Article 24.1). In addition, this article contains references to the 'public interest' and contemplates the adoption of 'repressive measures' to guarantee the preservation of the environment. These elements reflect a classic republican approach to citizenship – albeit one that emphasizes the 'active state' rather than the 'active citizen'. In contrast, the 1996 South African Constitution states that: 'Everyone has the right to an environment that is not harmful to their health or well-being' (Article 24). The text makes no explicit mention of duties, either of the state or of the citizens, although the protection of environmental rights is the implicit responsibility of the government 'through reasonable legislative and other measures' (Article 24). The discourse of rights and the emphasis on reasonableness denote a liberal articulation of environmental citizenship. This is also evidenced by the inclusion of environmental rights (as a fourth layer of individual rights) alongside civil, political and social rights, in a comprehensive Bill of Rights embedded in the Constitution.

The incorporation of environmental provisions, in the vast majority of constitutions, does not pose significant challenges to the current articulation and practice of citizenship, at least not challenges that cannot be easily accommodated into the classical dominant conceptions of citizenship, liberal and republican. There are, however, some exceptions that deserve special attention, because they represent more expansive, pluralist and ambitious articulations of environmental citizenship, including aspects that invoke notions of ecological or, perhaps more appropriately, ecocentric citizenship. The most interesting exceptions come from two countries of the Global South: Brazil and Ecuador.

The Brazilian Constitution: indigenous environmental citizenship

The first expansive constitutional text in matters regarding the protection of the environment was the 1988 Brazilian Constitution. The environmental provisions are contained in a single article that opens with the following statement:

> Everyone is entitled to an ecologically balanced environment, which is an asset of everyday use to the common man and essential to a healthy quality of life; this imposes a duty on the government and the community to protect and preserve it for the present and future generations.
>
> (Article 225)

The article includes six additional paragraphs, the first of which lists seven specific instructions for the government designed to guarantee the right to an ecologically balanced environment contained in the opening statement. The provisions reflect a wide range of environmental issues: the protection against pollution and the installation of nuclear reactors; the protection of species and national parks; the protection of ecosystems and animal rights; and the protection of the genetic patrimony. Some of the provisions challenge traditional notions of citizenship, most notably the protection of ecosystems and animal rights. However, in general, the provisions display a balance between liberal and republican conceptions of environmental citizenship, between the right to a healthy and ecologically balanced environment and the corresponding duty of the government and the community to protect and preserve that environment.

The connection between citizenship and the environment in the Brazilian Constitution is not exhausted by Article 225. The constitution includes a series of articles on the rights of indigenous peoples, granting them cultural and land rights, which have direct implications for environmental citizenship. The most important of these is the provision stating that indigenous peoples must be consulted in matters relating to their lands (Article 231). In general terms, the combination of indigenous rights and environmental rights enables a form of indigenous environmental citizenship to emerge – in the sense, or at least to the extent, that these provisions grant indigenous peoples the opportunity to shape environmental policies in ways that reflect their distinctive cultural values and cosmologies, including their understanding of the relationship between humans and nature. These provisions have significant environmental implications on the ground, largely derived from the fact that indigenous peoples are at the forefront of the environmental struggle for the future of the Amazon (Figure 5.1).

Yet, indigenous citizenship, environmental or otherwise, is no guarantee of green outcomes or effective environmental protections. Indigenous peoples are not green by definition – even if they often have a better track record than non-indigenous peoples at protecting the environment, at least in Brazil. More importantly, the developmental agenda still trumps environmental protections (and indigenous rights) in most countries of the Global South, not least in the case of Brazil. Here, the potential and limitations of constitutional environmental protections (even when coupled with indigenous rights) can be exemplified by the current struggle to stop the construction of the Belo Monte dam in the Amazon (Cao 2014). Despite the persistent and widespread environmental and indigenous opposition to the project, the government continues to push

Figure 5.1 *Munduruku Indians meet with Brazilian minister Carvalho to discuss indigenous rights*

Source: Agência Brasil. Photographer: Antônio Cruz.

ahead, illustrating how developmental objectives almost invariably override environmental concerns in the Global South. Similar dynamics can occur even in the presence of ecocentric constitutional provisions, as is the case in Ecuador.

The Ecuadorian Constitution: the rights of nature

The most radical constitutional conception of environmental citizenship can be found in the 2008 Ecuadorian Constitution. This is a long and detailed text, consisting of 494 articles. The document contains seven sections (under Chapter 2: Biodiversity and Natural Resources) on matters related to the environment, covering a wide range of issues: from climate change to bike lanes, from state regulation to citizen participation, from genetic technology to food sovereignty (Articles 395–415). The constitution has also a section on 'rights to good living' that include the 'right to water' (Article 12) and the 'right to food' (Article 13). In addition, the text recognizes the right to 'a healthy and ecologically balanced environment' (Article 14). However, the radical break comes in Chapter 7: Rights of Nature (Articles 71–74).

The Ecuadorian Constitution treats nature as a subject of rights. Specifically, the text grants nature 'the right to integral respect for its existence and for the maintenance and regeneration of its life cycles, structure, functions and evolutionary processes' (Article 71), and 'the right to be restored' (Article 72). Importantly, and to overcome the fact that nature cannot represent itself, the text grants legal standing to any person, community and nation that wishes to call upon public authorities to enforce the rights of nature (Article 71). These provisions challenge the traditional understanding of citizenship, extending its reach beyond human relations to encompass human–nature relations, and coming closer than any other constitutional text to the concept of the 'natural contract' theorized by Michel Serrès. Indeed, the 'rights of nature' contained in the Ecuadorian Constitution effectively compel everyone (not only the citizens of Ecuador) to live in harmony with and protect Mother Earth.

The text in general, and the rights of nature in particular, are underpinned by the indigenous notion of '*kawsay sumak*' ('*buen vivir*' in Spanish and 'good living' in English), a conception of life that seeks to break with the classical notions of development centred on perpetual economic growth, linear progress and anthropocentrism (Acosta 2008). At its core, there is a communitarian and austere ethos that makes 'good living' closer to the notion of 'a good life' than that of 'the good life'. The concept, derived from the indigenous peoples of the Andean region, gives centre stage to 'Pacha Mama' (Mother Earth), but ultimately promotes the dissolution of the society–nature dualism: 'Nature becomes part of the social world, and the polis is widened to include the nonhuman Nature/Actors e.g. animals, plants or ecosystems' (Gudynas 2011a: 445). Similar concepts inform the 2009 Bolivian Constitution, which also includes references to Mother Earth, albeit only in its Preamble. The text contemplates the right to a healthy and ecologically balanced environment, but does not grant rights to nature. However, in 2010 the Bolivian parliament passed a Law of the Rights of Mother Earth contemplating seven rights, namely, the rights to: life, diversity of life, water, clean air, equilibrium, restoration and freedom from contamination. Similar to the Ecuadorian Constitution, this law (now part of a larger legal framework) grants legal personhood to nature and allows citizens to sue in defence of 'Mother Earth'.

The radical nature of these provisions must be placed in its material and political context if we wish to appreciate both their potential and their limitations. It is no coincidence that these texts have been elaborated and approved under the New Left governments that govern in much of Latin America. Both the Ecuadorian and the Bolivian constitutions are expressions of a postcolonial rejection of capitalism and the neoliberal approach to development that has become popular across Latin America. These texts are political manifestos against the exploitation, by transnational capitalism, of impoverished and developing nations that rely on their natural resources for the provision of basic social services. In this sense, the rights of nature contemplated in the Ecuadorian Constitution can be seen as a political tool to protect Ecuador against future governments that might take a neoliberal approach to the management of the country's natural resources. This is clearly the objective behind constitutional provisions limiting the profits that can be earned by companies producing natural resources (e.g. Article 408). Yet, at the same time, the constitution allows the state to exploit otherwise protected natural resources, and to adopt otherwise prohibited technologies, provided the 'national interest' is at stake (e.g. Article 407). Thus, despite all the talk about the rights of nature, the radical potential of the text and its leftist ideological foundations, the Ecuadorian Constitution ultimately enables the business-as-usual extraction of natural resources, albeit by the state.

The tension between environment and development, the rights of nature and human rights, can be seen in the current dispute over the extraction of oil from the Yasuní National Park, under the Yasuní ITT Initiative (see Box 5.2). This case illustrates the extent to which development and anthropocentrism can shape environmental outcomes even in cases where the presence of constitutional provisions granting 'rights to nature' might suggest an ecocentric approach to citizenship, a contract with nature. Indeed, for all the talk of the rights of nature, the logic of the Yasuní ITT Initiative is economic, and reflects a view of nature as resource. Thus, for example, the compensation demanded by the Ecuadorian state is based on the potential loss of revenue from oil extraction, not on the potential impact of oil extraction on the ecosystem, on nature (Gudynas 2011b). The initiative also reveals the underlying tension – also noticeable in the constitution – between the need to protect the rights of nature and the need to fund social programmes in order to secure people's social rights. The outcome of this initiative will be a good indication of the extent to which the constitutional protection of the rights of nature will have radical, significant, marginal or no repercussions when those rights come into conflict with other state priorities, such as economic development and the provision of human services, particularly when a country has no other major source of income than its natural resources.

Box 5.2

The Yasuní ITT Initiative

The 2007 Yasuní ITT Initiative is an innovative scheme to address concerns about climate change by leaving fossil fuels in the ground in exchange for partial compensation for the loss of revenue that would be derived from their exploitation. The initiative called for $3.6 billion from the international community – equal to 50 per cent of the estimated value of the oil under the Yasuní National Park – in exchange for leaving the 850 barrels worth of crude in the ground. The principle behind the initiative is that the costs associated with addressing our shared environmental concerns, particularly regarding climate change, should be shared by all, in this case by the rest of the world going halves with Ecuador. Ecuador would sacrifice half of its revenue from the value of the oil and the rest of the world would contribute the other half.

However, after six years, with $300 million pledged and only $13 million deposited (less than 1 per cent of the target sum), President Rafael Correa announced that drilling could begin as early as 2016. The announcement came as a major blow to campaigners who had backed the initiative as a pioneering alternative to the relentless extraction of fossil fuels both globally and in the Amazon. The president has promised 'responsible exploitation', but few trust that he (or anyone else) can deliver on that promise, let alone extract oil without causing significant damage to an area of outstanding natural wealth and biodiversity, in which one hectare contains as many species as the entire continent of North America. The remote area is also home to two indigenous groups that have chosen to live in voluntary isolation and that now face coming into contact with those drilling the oil of the Yasuní National Park.

Resources

Ecuador Yasuní ITT Trust Fund. Available at: http://mptf.undp.org/yasuni
Yasuní ITT documentary (uploaded 21 October 2009). Duration 6.38 mins. Available at: www.youtube.com/watch?v=cr3oep32SzE

The future: environmental human rights or the rights of Mother Nature?

The two contrasting governmental approaches to environmental citizenship we have seen in this section – the mainstream and the radical – can also be found at the international level, even if only in the form of projects and proposals. The mainstream approach is illustrated by the Draft Declaration of Principles on Human Rights and the Environment (Draft Declaration) produced in May 1994 by an international group of experts who met at the UN seat in Geneva. The radical approach is illustrated by the Universal Declaration of the Rights of Mother Earth, proclaimed at the World People's Conference on Climate Change and the Rights of Mother Earth, held in Cochabamba, Bolivia, in April 2010 (Figure 5.2).

The Draft Declaration spells out the content of 'the right to a secure, healthy and ecologically sound environment' (Principle 2). The document contemplates substantive and procedural environmental rights and explains the duties of individuals, states, international organizations and transnational corporations (Aguilar and Popović 1994). Its content reflects a conception of rights as indivisible and interdependent, often bringing together the right to a satisfactory environment and the right to development (e.g. the right to housing in an ecologically sound environment, the right to a safe and healthy working and living environment, and the right to environmental and humans rights education). Since then, the international environmental rights agenda has

Figure 5.2 *World People's Conference on Climate Change and the Rights of Mother Earth (2010)*

Source: Wikimedia Commons. Photographer: Kris Krüg.

been somewhat dormant, with the notable exception of the ratification of the Aarhus Convention (explored later in this chapter).

In contrast to this relative slumber, the government of Bolivia has been at the centre of the political project to advance globally the rights of nature, most notably as the host nation of the gathering of civil society and governments in Cochabamba. The event was a response to the climate talks of the 2009 Copenhagen United Nations Conference on Climate Change, and an attempt to launch a counter-movement to fight against climate change from an anti-capitalist platform that underlines the New Left in Latin America. The conference included a World People's Referendum on Climate Change, the establishment of a Climate Justice Tribunal, and the formulation of a Universal Declaration on the Rights of Mother Earth, which opens with the statement: 'We, the peoples and nations of Earth'. The declaration includes a lengthy list of rights, classified under three articles: Mother Earth (Article 1); Inherent rights of Mother Earth (Article 2); and Obligations of human beings to Mother Earth (Article 3).

The 'Rights of Nature' included in the Ecuadorian Constitution and the Universal Declaration of the Rights of Mother Earth probably sound aspirational and perhaps even implausible to most of us – especially when compared with more conventional initiatives such as the Draft Declaration. Indeed, environmental rights in general and the rights of nature in particular will remain only aspirations (not rights) until they are properly enforceable – and until states are able (and willing) to resist the economic pressures that push them to override environmental provisions. Yet we must remember that previous revolutionary texts, such as the American Bill of Rights (1789) and the French Declaration of the Rights of Man and the Citizen (1789), were also limited by their historical context, and would have been considered little more that aspirational texts at the time, particularly by those who were or felt excluded, such as slaves, women and the poor, but in time the impact of their ideals came to be realized. Similarly, the Universal Declaration of Human Rights is still widely seen as an aspirational text, and yet its positive impact on the protection of human rights is unquestionable.

Granting legal rights to nature may sound far-fetched at present, but as Christopher Stone stated in his classic article 'Should Trees Have Standing?' (1972), each extension of rights to some new entity has been 'a bit unthinkable' (1972: 455). Stone notes: 'Each time there is a movement to confer rights onto some new "entity", the proposal is bound to sound odd or frightening or laughable' (1972: 455). This is because until the entity in question is recognized has having rights, 'we cannot see it as anything but a thing for the use of "us" – those who are holding rights at the time' (Stone 1972: 455). This is currently true for nature (and to a lesser extent, but still, for animals) as it has been for slaves, women and children at different times in history. Thus, whilst we might not see the full realization of the rights of nature in our lifetime, we are witnessing the emergence of a radically different way to conceptualize citizenship, with potentially profound practical implications in years to come.

The question, perhaps, is not whether this will come to pass, but what kind of transformation this would bring if it did. After all, nature cannot represent itself and thus the question of who will stand for nature has crucial implications. If the state becomes the key agent, these changes can create the ultimate Leviathan, a state that can regulate all life. If experts and scientists become the dominant agent, we might see a shift of power from citizens to scientists – with the potential for a scientific technocracy. If, on the other hand, citizens become the key agents, we might see the emergence of a natural democracy. In any case, we can expect to see a continuing and increasingly heated battle to define (and defend) the rights of nature vis-à-vis human rights – a battle fought, as always and inevitably, by and between human beings.

Environmental governance: green(ing) neoliberal citizenship

The governance approach to environmental matters is now the dominant paradigm, especially at the international level. The regulatory multilateral frameworks established and coordinated by the United Nations Environment Programme (UNEP), and regional political institutions such as the European Union, have a powerful normative influence, and often shape national legislative agendas. It is important to remember, however, that governance is not the exclusive domain of governments. Indeed, by its very nature, this form of governing leaves plenty of room for non-state actors not just to collaborate, but also to take the lead in governing. Thus, we should not be surprised to find out that the governance approach to the environment has also been embraced by corporations (as we shall see in Chapter 6) and NGOs, particularly large international NGOs, most notably the International Union for Conservation of Nature (IUCN). In this section we will explore the impact of environmental governance on environmental citizenship, paying particular attention to the increasing governmental recruitment of citizens.

Cooperation and participation: towards neoliberal citizenship

The earliest adoption of the governance approach towards the environment began at the national level, and is best illustrated by the case of Germany. Ever since the environmental movement gathered strength in the 1970s, the German government has been one of the most active in creating a robust legislative and regulatory framework to shape industrial and social environmental behaviour (Winter 2012). One of the earliest and most significant developments was the establishment of the Emergency Program for Environmental Protection (1970). This programme – designed to address the level of air, water and noise pollution, deal with waste disposal and strengthen the protection of nature – was based on three principles: prevention, polluter-pays and cooperation. The prevention principle aims at avoiding pollution and environmental risks before they occur. The polluter-pays principle assigns the costs of pollution to the polluter who is responsible – with the government bearing the cost only when no distinct polluter can be identified. The cooperation principle states that environmental protection is a task that must be shared equally by the government, the citizenry and the business sector. The cooperation principle is of particular significance here. In addition to illustrating the environmental governance approach to environmental policy, the integration of corporations announced the shift towards market economics and neoliberal citizenship that would come to dominate in the 1990s.

The same double trend towards environmental governance and neoliberal citizenship can be appreciated in the environmental policy of the European Union (EU). The most explicit engagement with the language of citizenship in EU policy is the Convention on Access to Information, Public Participation in Decision-Making and Access to Justice in Environmental Matters (1998), commonly known as the Aarhus Convention. The Convention lays down the basic rules to promote citizen involvement in environmental matters and enforcement of environmental law, and provides a citizenship framework on environmental policy across the EU. The text establishes three basic rights for European citizens, reflected in the official title of the declaration: access to information, public participation in decision-making and access to justice in environmental matters. Insofar as it enhances citizen participation and public accountability, the Convention has been rightly referred to as 'a modest contribution to environmental democracy' (Ebbesson 2011). In essence, the Convention is intended to ensure procedural and participatory rights in environmental matters – making this a useful instrument for the exercise of liberal environmental citizenship. However, the text can also be seen as part of the gradual transfer of responsibility from governments to citizens regarding the monitoring of environmental matters.

The incorporation of citizens as stakeholders in the process of governing the environment goes beyond general mechanisms for cooperation and legal participation, and includes hands-on participation alongside governments and NGOs. This is particularly noticeable in the area of nature conservation. Thus, for example, Rebecca Ellis and Claire Waterton (2004) document shifts in biodiversity policy in the United Kingdom that include attempts to enrol 'volunteer naturalists' and lay citizens into biodiversity action planning. Their study reveals how elite actors in UK biodiversity policy are trying to harness the knowledge and time of civil society as part of an official endeavour to know and represent biodiversity. Under the UK Biodiversity Plan, these policy actors have formed a partnership consisting of the statutory agency for nature conservation, English Nature, and the UK Biodiversity Group at the Natural History Museum in London. This initiative, based on the exchange of knowledge of nature amongst the different communities involved, is an example of 'environmental citizenship in the making' (Ellis and Waterton 2004). The inclusion of volunteers into nature conservation projects has also been embraced by environmental NGOs (Lorimer 2010).

The growing emphasis on citizen participation in governing the environment, within the broader push for self-regulation and individual responsibility, suggests a close relationship between environmental governance and environmental governmentality, and illustrates the gradual shift from liberal to neoliberal conceptions of environmental citizenship. This trend takes on a global dimension under the auspices of the UNEP, the coordinating agency for global environmental governance since 1972.

Global neoliberal environmental citizenship

The UNEP's formulation of its central operational concept, sustainable development, reveals an increasingly direct and explicit engagement with the notion of environmental citizenship, as well as a gradual shift towards neoliberal environmental citizenship. The early declarations that frame its global agenda reflect a liberal conception of environmental citizenship, focused on the right to a healthy environment for present and future generations. Thus, Principle I of the 1972 Stockholm Declaration states:

> Man has the fundamental right to freedom, equality and adequate conditions of life, in an environment of a quality that permits a life of dignity and well-being, and he bears a solemn responsibility to protect and improve the environment for present and future generations.
>
> (UNEP 1972)

The same sentiments are at the heart of the declaration that followed the 1992 UN Conference on Environment and Development, the so-called Earth Summit, held in Rio de Janeiro. Principle I of the Rio Declaration states: 'Human beings are at the center of concerns for sustainable development. They are entitled to a healthy and productive life in harmony with nature' (UNEP 1992a).

The UNEP's deployment of the language of citizenship has become more frequent and explicit in recent times. Thus, for example, the organization has funded initiatives to promote global environmental citizenship, such as the project on 'Global Environmental Citizenship' for Latin America and the Caribbean, launched in November 2001. This three-year project was funded by the UNEP and carried out in collaboration with six Latin American networks, consisting of parliamentarians, consumer organizations, local authorities, educators, radio broadcasters and religious leaders, together with the environmental authorities of Argentina, Chile, Costa Rica, Cuba, Ecuador, Mexico and Peru (UNEP 2001). The range of participant networks is an excellent illustration of the collaborative character of environmental governance. The project included the development of demonstration activities aimed at establishing national strategies on environmental citizenship and the preparation of specific proposals to strengthen the legislative framework in the region on those same issues. In addition, the project was expected to generate public awareness, increase levels of understanding of global environmental issues and mobilize support in the countries of the region for the objectives of the Global Environmental Facility (GEF). The adoption of the language of environmental citizenship in a project soaked in the language of neoliberal governance (e.g. partnerships and stakeholders) illustrates the extent to which this notion of citizenship is being deployed by global authorities as part of their managerial, neoliberal agenda, with the UNEP at the front of the pack.

However, despite being a global regime, the UNEP is very aware of the crucial significance of the local political space for the successful implementation of their environmental agenda. Indeed, the UNEP has explicitly embraced the motto often associated with environmental activism: 'Think globally, act locally'. This is clear in its signature document of the Earth Summit, and still its main operational framework, Agenda 21. Since the start, the UNEP has recognized the need for local action and involvement in the design and implementation of Agenda 21. Thus, for example, Chapter 28 requests the involvement of local authorities in the development and implementation of local action programmes in the form of 'a local Agenda 21' (UNEP 1992b). The invocation of the language of citizenship coupled with the emphasis on the local is also noticeable in UNEP's 2002 Annual Report. In the section entitled 'Environmental Citizenship Program', the report states that it is the duty of citizens 'to help protect and conserve that part of the local ecosystem where he or she belongs to or is a part of and to participate actively in local environmental affairs in cooperation with government and others' (UNEP 2002).

The incorporation of the general citizenry into environmental governance began as an invitation by governments in response to environmental activism, and was perceived by citizens in general and activists in particular as an opportunity to influence policy directly. However, the initial enthusiasm over the prospect of participation in the decision-making and implementation of the sustainable development agenda in local communities has since evaporated, as citizens realize that many of their ideas are being disregarded or put aside 'for the time being', and that control remains largely in the hands of politicians and administrators (Feitchtinger and Pregernig 2005). The growing sense is that the experts make decisions in advance and citizen participation is simply enlisted and managed. Moreover, commercial agents have increasingly moved in as mediators, promoting 'consensus- and entertainment-oriented varieties of citizen involvement' (Læssøe 2007: 247). The result is a post-ecologist politics that 'easily brushes over fundamental tensions and conflicts and ends up with the self-deceptive simulation of sustainable development' (Læssøe 2007: 247). This post-ecologist turn implies a narrowing of citizenship, from actions defined by political engagement and collective empowerment to 'an approach that avoids politicization and promotes small technical fixes' (Læssøe 2007: 246). In other words, environmental governance has produced a form of post-ecologist citizenship that undermines notions of ecological citizenship before citizens have had the opportunity to develop and fully engage with them – at least in the context of relations between citizens and governments. Furthermore, some of these post-ecologist dynamics are suggestive of techniques of governmentality, as we shall see in the next and final section.

Environmentality: governing through green citizens

The hegemony of the neoliberal mentality, with its aversion to government regulation, has translated into the adoption of more indirect and diffuse techniques for the governing of the environment. There is a noticeable shift from governing through direct regulation to governing through citizens by means of pedagogical and psychological techniques. This trend is particularly pronounced in the affluent countries of the Global North, but is being extended to the Global South through agencies such as the UNEP. This approach is best defined as 'environmental governmentality' or 'environmentality' (Agrawal 2005). This form of governing the environment involves 'the use of social-engineering techniques to get the attention of the population to focus on specific environmental issues and to instill – in a non-openly coercive manner – new environmental conducts' (Darier 1996: 594). In the present context, environmentality is heavily infused in neoliberal values, ideals and priorities. The ascendancy of neoliberal environmentality has coincided with the increasing adoption of environmental citizenship as a strategy to address environmental issues – suggesting a functional (if not constitutive) affinity between the two. In this section, we will explore the most significant points of intersection between environmentality and environmental citizenship, beginning with the first explicit deployment of the concept of environmental citizenship by a governmental agency, namely, Environment Canada.

Environment Canada: governing through literate citizens

In 1990, the Canadian government launched its Green Plan, vowing to make Canada the most environmentally friendly country in the industrial world. The plan represented a shift from deploying regulation and the law to deploying education and awareness as the main tools for achieving environmental sustainability in Canada (Darier 1996). The government argued that state action on its own could not achieve its sustainability goals and decided to mobilize the entire population. In addition, the government took the view that self-regulation and voluntary action were the most effective way to achieve enduring results (Darier 1996: 595). In order to pursue this agenda, the government deployed the notion of environmental citizenship, mainly through its environmental agency, Environment Canada. The agency defined environmental citizenship as follows:

> Environmental citizenship is a personal commitment to learning more about the environment and to taking responsible environmental action. Environmental citizenship encourages individuals, communities and organizations to think about the environmental rights and responsibilities we all have as residents of planet Earth. Environmental citizenship means caring for the Earth and caring for Canada.
>
> (Environment Canada 1993: 1)

This definition contains several aspects that have come to be typically associated with environmental citizenship, namely, the expanded conception of political community and the emphasis on individual responsibilities. The expanded conception of political community is reflected in the reference to residents of planet Earth – although the final twist indicates that this is a complementary community, an addition to rather than a replacement of the national community. The emphasis on individual responsibilities is reflected in the contrast between the single reference to rights and the four duty-related references: commitment to learning; taking responsible action; thinking about rights and responsibilities as residents of the planet; and caring for the planet and for Canada. Yet, despite the emphasis on duties, citizenship is essentially framed as voluntary and personal. This notion of environmental citizenship was expanded on the agency's website.

Environment Canada – in content no longer available on its website – told the citizen that:

> Each of us has an effect on the environment every day; the key is to make this impact a positive one. We must all take responsibility for our own actions, whether as individuals, or as members of a community or an organization. Let's work together and become good Environmental Citizens. If you don't, who will?

Environment Canada then informed the public of ways in which they could make a difference and be active environmental citizens. These included instructions such as: turning off the tap when brushing your teeth, or washing your face; walking, riding your bike, car sharing or using public transport when possible; and shopping at second-hand stores and garage sales, instead of purchasing brand new items. These examples illustrate the privatization and individualization of responsibility for the environment that has become commonplace in recent times. Citizens are taught and called upon to be self-disciplined in their everyday life, to be good citizens, increasingly in their capacity as consumer-citizens. This approach is not limited to governments. NGOs now also advocate and promote this kind of informed, self-disciplined approach to individualized environmental responsibility (e.g. WWF Malaysia and Partners 2008).

The most salient aspect of the environmentality approach taken by the government of Canada is its pedagogical dimension. The centrality of environmental education in the formation and enactment of environmental citizenship reveals the process of governmentality at work, that is, the process of governing through citizens rather than through regulations. The pedagogical character of this approach is perfectly illustrated in the 'Canadian Environmental Citizenship Program' associated with the Green Plan, expertly analyzed by Eric Darier (1996). Darier notes how the six objectives included in the programme are mostly educational, i.e. related to learning and understanding (see Box 5.3). In addition, the plan contemplated a series of programmes 'designed to encourage the introduction of regular environmental exercises or drills' (Darier 1996: 599). In essence, the programme constitutes a clear attempt to discipline the population by instilling new norms of environmental conduct that construct a new subjectivity based on the concept of environmental citizenship. Darier also notes the temporal framing of the project – presented explicitly as 'a plan for life'– and warns about the potential risk to democracy and freedom that can arise from the depth and breadth of government programmes that deploy environmental citizenship as 'life politics' (1996: 597). These schemes aimed at governing through citizens, can effectively lead to transforming citizens into subjects, instruments of an agenda over which they have little, if any, control.

The conception of environmental citizenship in Environment Canada has been criticized on several grounds, not least because it depoliticizes environmental issues by removing both discussion and action from the public sphere. In theory, environmental citizenship reflects the actions of informed and concerned citizens. However, the techniques used to instil values and 'nudge' citizens towards a particular course of action, without their direct involvement in the process of deliberation, let alone the discussion of priorities and the setting of the agenda, are clearly problematic. The condition under which citizens act is far from the ideal of the autonomous citizen and more reflective of a condition of 'regulated autonomy' (Rose 1996). In other words, whilst in theory their actions reflect the free actions of concerned citizens (i.e. there is

---

Box 5.3

The Canadian Environmental Citizenship Program

The six aspects included in the Canadian Environmental Citizenship Program contemplated in the Green Plan (1990) are:

- *campaigns* designed to enhance environmental *awareness* and promote public participation;

- the development of *learning* materials and programmes designed to promote *understanding* and motivate informed decision-making at all levels of society;

- the development of specialized *campaigns* on such issues as climate change, waste management, the prevention of water pollution, and water conservation;

- an exchange of environmental *learning* resources on a national and international basis;

- the development and implementation of environmental action and training plans appropriate for specific audiences; and

- support for partnership activities designed to enhance general environmental *awareness* and increase *understanding* of specific issues.

(Darier 1996: 595–596, emphasis added)

---

no overt coercion), there is a level of manipulation (more or less subtle, diffuse and intrusive) that, without being overtly coercive, has coercion-like effects. In addition, this approach has also been criticized because the government avoids taking (or even discussing) difficult political decisions, and thus effectively transfers responsibility for the environment to individuals.

The centrality of environmental education for environmental citizenship, far from being accidental, reflects the pedagogical nature of this new form of governing the environment called environmentality. The strong link between environmental governmentality and environmental citizenship raises some significant issues regarding the uncritical embrace of the discourse of environmental citizenship, particularly as deployed by political authorities and governance regimes. The most important of these concerns is the extent to which citizens remain autonomous beings or are transformed into regulated and instrumental subjects of environmental policies. In any case, this way of governing the environment through citizens necessitates the manufacturing of individuals into environmentally literate and responsible citizens. These reflections and conclusions should be kept in mind (and will be invoked again) when we explore environmental education and ecological pedagogies in Chapter 7.

Sustainable consumption: governing through green consumers

The convergence of neoliberal governmentality and environmental citizenship is closely related to the emergence of sustainable consumption as a fundamental articulation of environmental citizenship. This approach to governing the environment through green consumers (or consumer-citizens) has been embraced prominently in the UK. The most recent illustration of this strategy has been the Green Deal, launched in 2012. The Green Deal is an initiative that provides loans for energy-saving home upgrades. Customers borrow the money to make the upgrades, and pay back the loan through their energy bills. The policy is based on market-driven, public-private approaches in line with the government's policy of relying on markets as the mechanism for delivering environmental policies and outcomes. But this is only the latest initiative in a history of programmes centred on changing individual behaviour rather than making structural reforms when it comes to promoting sustainability and sustainable development in the UK.

The evolution of the governmental approaches to sustainable development is well mapped and analyzed by Stewart Barr (2008). His analysis of the government's environmental campaigns reveals an anthropocentric 'light green' approach to sustainable development that embraces market forces at the expense of direct government regulation (see Box 5.4). The campaigns have shifted progressively from a centralized top-down approach, based on the assumption that people would comply with government advice once they became aware of the issues and of what to do, to a distributed bottom-up approach that tries to encourage citizens to get involved and propose their own green initiatives (Barr 2008). Despite some concessions to notions of community in recent years, the campaigns are invariably aimed at altering individual behaviour, and engage the citizen, first and foremost, as a consumer. The net effect of these campaigns is the creation of a green consumer-citizen through which the government can conduct the governing of the environment.

Box 5.4

Sustainable development in the UK

Stewart Barr has examined four sequential approaches initiated by the central government in the UK to change behaviour towards the environment. The four campaigns are:

- *Helping the Earth Begins at Home* (1991). This campaign focused on the impact of carbon emissions from personal energy use in the home on the global environment, most notably the greenhouse effect. The campaign equated saving energy with saving money and promoted market forces and self-regulation as the preferred road towards sustainability.

- *Going for Green* (1995). This campaign was a partnership between the central government and the private sector, with funding from major corporations. The campaign had five central messages for consumers: travel sensibly, prevent pollution, cut down waste, save energy and natural resources, and look after the local environment.

- *Are You Doing Your Bit?* (1999). This campaign introduced sustainable consumption as central to the sustainable development strategy, recommending small, incremental changes to individual lifestyles, particularly in the home (e.g. turning off lights in unused rooms). The campaign was delivered through multimedia and by well-known celebrities.

- *Together We Can* (2005). This campaign, despite its more community-oriented title, retained the focus on individual behaviour, only this time with greater emphasis on positive and inspirational messages rather than on the narratives of fear and concern of earlier campaigns.

(Adapted from Barr 2008: 86–93)

Addressing environmental matters through consumer-citizens is also evident at the EU level. This approach is well illustrated in the campaign against climate change entitled: 'You Control Climate Change' (2006). The campaign tells consumers to: 'Turn down. Switch off. Recycle. Walk'. The objective is to get citizens to make small changes in their daily routines to reduce greenhouse gas emissions and tackle global warming. The campaign, with its tips, greenhouse gas calculators, games and information, effectively creates a 'climate citizen' in the EU. This citizen is defined as an individual – or, at most, a household – that performs small, private sphere activities in order to reduce greenhouse emissions. The focus is on individual action, with the great majority of

Figure 5.3 *Awareness raising activities across Europe*

Source: European Environment Agency.

the advice and tips regarding patterns of private behaviour, most of them related to the home (e.g. recycling and switching off lights) and/or home-related matters (e.g. walking to the local store and taking own bags to go shopping). The campaign also targets students, who are encouraged to sign a pledge to reduce their carbon dioxide emissions. This notion of individuals making a public commitment to reducing greenhouse gas emissions has since extended to different locations (Upham 2012).

Whilst such changes in personal behaviour are laudable in and of themselves, the emphasis on small-scale initiatives and individual behaviour occludes large-scale structural issues underlying climate change, such as emissions from coal-based electricity production (Figure 5.4). In addition, this approach diverts attention from matters of environmental justice, such as climate debt, that would require addressing the disproportionate historical contributions to the current problem of climate change by the countries of the Global North and those of the Global South. Moreover, in a context where the wider causes of climate change are felt to be beyond individual control, some authors argue that

Figure 5.4 'A good start'

Source: Cartoon courtesy of Andrew Weldon.

individualized responsibilization can lead to resignation and acceptance rather than action, effectively resulting in 'deresponsibilization' (Butler 2010: 183). Last but not least, there are no tips or advice given to citizens about engaging in climate politics. In other words, there is no representation or interpellation of the citizen as a political actor, only as an economic being.

The governmental push to promote green consumer-citizens has become part of the global environmental strategy of the UNEP, as illustrated by its campaign 'Shopping for a Better World' (UNEP 2003). The project focuses on marketing 'attractive' or 'desirable' lifestyles as the key to selling environmentally friendly products. The aim is to show how sustainable lifestyles can be fashionable. In the words of UNEP Executive Director, Klaus Toepfer: 'what can be more modern, more fashionable, than caring about our planet?' (UNEP 2003). The programme targets forms of inappropriate consumption, not overconsumption. 'Sustainable consumption – reads the press release – is not about consuming less, it is about consuming differently, consuming efficiently, and having an improved quality of life' (UNEP 2003). Citizens (as consumers) are expected to exercise self-control regarding what and how they consume, not how much they consume. The main message seems to be that 'there is no harm in consuming if each individual pays attention' (Rumpala 2011: 695). Once again, the focus on individual consumer behaviour diverts attention from unjust and environmentally unsustainable structural patterns of consumption, particularly between the Global North and the Global South. The same logic informs a recent scheme that explicitly invokes the language of citizenship and is designed to 'encourage demand for sustainable tourism products and services': the Green Passport (UNEP 2014).

The environmental citizen produced by these and similar government initiatives is someone who has internalized information about environmental problems (a literate citizen), creating a sense of personal responsibility and duty that is then expressed mainly through consumption (a consumer-citizen). This disciplining of private citizens into green consumer practices which do not challenge consumer capitalism can be described as neoliberal environmental governmentality or neoliberal environmentality. These schemes operate by deeply inscribing self-monitoring and self-control into citizens which – in the absence of inclusive and transparent deliberations– perpetuate social control and reinscribe hierarchies into the seemingly democratic operations of government (Davies 2012). This trend to govern through green citizens, at the expense of government regulation and structural reforms, is partly driven by the increasing influence of corporations in the governing of the environment, something for which they are increasingly relying on their own articulation of environmental citizenship, as we shall see in the next chapter.

Summary points

- There has been increased government action on environmental issues since the 1970s. This has taken several forms: standard legislation, constitutional provisions and, increasingly, pedagogical programmes and public campaigns.
- There has been a noticeable global shift from environmental regulation to environmental management in the governing of the environment. This invokes the concepts of governance and governmentality, which, together with traditional government, form 'the three govs'.
- The constitutionalization of the environment by and large combines liberal and republican conceptions of citizenship, but some countries

have embraced or are contemplating provisions that resemble the 'natural contract' in the form of 'rights of nature'.

- All protections are limited by the material and political context in which they operate. These contexts have a powerful influence on the capacity of governments to implement their conceptions of environmental citizenship, especially the more radical ones.
- The growing emphasis on self-regulation and the individualization of responsibility suggests a gradual shift from liberal to neoliberal conceptions of environmental citizenship.
- Governments have adopted environmental citizenship as part of a more diffuse form of governing societies and resolving environmental issues, known as environmental governmentality or environmentality.
- Neoliberal environmental citizenship promotes the individualization of responsibility through sustainable consumption, whilst eliding the structural causes of global environmental deterioration and patterns of inequality between the Global North and the Global South.

Selected readings

Bevir, M. (2011) 'Governance and Governmentality after Neoliberalism'. *Policy and Politics*, 39(4): 457–471.

Blühdorn, I. (2007) 'Sustaining the Unsustainable: Symbolic Politics and the Politics of Simulation'. *Environmental Politics*, 16(2): 251–275.

Darier, E. (1996) 'Environmental Governmentality: The Case of Canada's Green Plan'. *Environmental Politics*, 5(4): 585–606.

Environment Canada (1993) 'From the Mountains to the Sea: A Journey in Environmental Citizenship'. Canada: Published by authority of the Minister of the Environment. Available at: http://infohouse.p2ric.org/ref/14/13416.pdf (accessed 23 July 2013).

European Commission (2007) 'Change: Turn Down. Switch Off. Recycle. Walk. Make a Pledge!' Luxembourg: Office for Official Publications of the European Communities. Available at: http://ec.europa.eu/clima/sites/campaign/pdf/toolkit_en.pdf (accessed 23 July 2013).

Evans, J. P. (2012) *Environmental Governance*. London: Routledge.

Gudynas, E. (2011) 'Buen Vivir: Today's Tomorrow'. *Development*, 54(4): 441–447.

Hayward, T. (2005) *Constitutional Environmental Rights*. Oxford: Oxford University Press.

Popović, N. (1996) 'In Pursuit of Environmental Human Rights: Commentary on the Draft Declaration of Principles on Human Rights and the Environment'. *Columbia Human Rights Law Review*, 27(3): 487–603.

Rumpala, Y. (2011) '"Sustainable Consumption" as a New Phase in a Governmentalization of Consumption'. *Theory and Society*, 40(6): 669–699.

Online resources

European Commission: *Environment for Europeans*. EU Magazine of the
Directorate General of the Environment. Available at: http://ec.europa.eu/
environment/news/efe/index_en.htm

European Commission: You Control Climate Change (uploaded 12 June 2007).
Duration 8.24 mins. Available at: www.youtube.com/watch?v=UB1OUtt0DoE

See also the series of short films engaging with teenagers from 21 EU
member states, where they present their tips for tackling climate change:
'Climate Change: You Control It!' Available at: www.youtube.com/
watch?v=6dj0GJXUOIc (uploaded 6 June 2009). Available at: www.youtube.
com/watch?v=_yjMCmUYvsc (uploaded 23 November 2011).

More videos on 'The Climate: You Control It!' at: www.climatechangeeu.com

European Environment Agency (EEA). Available at: www.eea.europa.eu/

European Environment Agency: *Planet Re:think* (2012) Dir. Eskil Hardt.
Duration 1.25 mins. Available at: www.eea.europa.eu/media/audiovisuals/
planet-re-think/view

Environmental Protection Agency (USA) Available at: www.epa.gov/

United Nations Environment Programme (UNEP). Available at: www.unep.org/

Universal Declaration of the Rights of Mother Earth (2010). Available at: www.
rightsofmotherearth.com/declaration/

UNEP: Green Passport: Holidays for a Living Planet (2014). Available at: www.
unep.fr/greenpassport/

Student activity

Explore the citizenship profile of a government or constitution.
The websites of national governments and international governmental
organizations contain a wealth of photos, publications and videos
about their work on environmental issues. Select a government agency,
ministry or department, and explore the materials on their website:
homepage, vision statements or stated priorities, specific policies,
official reports and video galleries. Discuss the relation of that content
with the basic elements of citizenship: membership, rights and duties
(individual and collective). What kinds of rights are defended, and
for whom? What kinds of duties are defended, and for whom? Does
the agency reflect moderate or radical conceptions of environmental
citizenship? Is its perspective anthropocentric or ecocentric? Do
you think the agency you have chosen reflects a liberal, republican,
cosmopolitan, feminist, ecological or other conception of environmental
citizenship? Or a mix?

Alternatively, you can explore the constitutional text of a country and address the same questions. In addition, you can also address the following question: to what extent are these constitutional changes little more than 'simulative politics' (Blühdorn 2007)? Are these provisions having any impact on the environmental predicament and/or the actions of environmental citizens in the relevant countries?

Discussion questions

- Does your country have a constitution? If so, does it contain environmental provisions? If so, do they resemble liberal or republican citizenship?
- Should nature have specific rights? If not, why not? If so, what should these rights be?
- Do you think there is a problem with the individualization and privatization of responsibility when it comes to environmental matters? Why?/Why not?
- Do you think sustainable consumption policies deflect attention from government responsibilities and structural issues? If so, in what ways?

6 ⬤ Environmental citizenship incorporated

> The superstores' green conversion is astonishing, wonderful, disorienting. If Tesco and Wal-Mart have become friends of the earth, are there any enemies left?
>
> George Monbiot (2007)

Corporations have transformed citizenship in general and environmental citizenship in particular from a partnership of two into a troika. They have incorporated the language of environmental citizenship into their identities to the point that even the most consumption-oriented corporations claim to be model environmental citizens, true friends of the Earth. Whilst not everyone is convinced about the sincerity or effectiveness of that embrace, what is certain is that this development raises the potential to radically transform the practice of citizenship, and even its very meaning. This chapter explores the impact corporations have on the formulation and articulation of environmental citizenship, with attention to the main debates and controversies surrounding the concept of corporate environmental citizenship (CEC). The chapter also examines a particular conception of corporate citizenship with significant connotations for the future of environmental citizenship, the so-called supply-chain citizenship, exemplified here by Walmart. But before we delve into the environmental dimension of corporate behaviour, we must explore the question of whether corporations are more like citizens or more like governments.

Corporations: citizens and/or governments?

Corporations are, first and foremost, private economic agents. But, unlike other business organizations, corporations are also legal persons, which makes them legal entities separate from their constituent members or shareholders. Their legal status enables them to own property, undertake contractual relationships, engage in economic transactions and operate as unitary agents with an increasingly broad range of rights. In this sense, their legal status is similar to that of an ordinary citizen, with rights (e.g. legal equality, due process) and responsibilities (e.g. law abidance, paying taxes). Typically, those rights have been limited to economic rights and the civil rights attached to the protection of contracts. However, in recent times, corporations have been granted protections that amount to political rights, particularly following the famous and controversial ruling of the US Supreme Court in the case of *Citizens United vs Federal Election Commission*, in 2010.

In its ruling, the US Supreme Court held that the First Amendment of the US Constitution (i.e. the right to free speech) prohibits the government from restricting political independent expenditures by corporations, associations or labour unions. The ruling did not overturn the federal ban on direct corporate contributions to candidate campaigns or political parties, but effectively grants corporations the right to contribute indirectly to election campaign funding. This ruling has been credited with the creation of large political action committees, better known as 'super PACs'. Officially known as 'independent-expenditure only committees', these legal entities can accept unlimited contributions for political purposes from individuals, corporations and unions, and consequently engage in unlimited political spending at any time, including during political campaigns. Irrespective of whether super PACs corrupt the political system or not, the Supreme Court decision effectively has granted corporations the right to participate in political campaigns as citizens in their own right – and has done so in the name of a civil right traditionally associated with human beings: the right to free speech.

The citizen-like character of corporations becomes more evident as we look at their increasing involvement in the governing of societies through the governance approach, alongside other citizens and their political representatives. This approach to governing has enabled and demanded that corporations take on greater responsibilities as citizens – something they have responded to by embracing the principles of corporate social responsibility (CSR), a concept we will explore in the next section. The addition of responsibilities to the aforementioned rights produces a much more complex profile of the corporation as citizen, one that increasingly resembles a fully rounded, multidimensional natural person (Crane *et al.* 2008a).

A great deal of anxiety has been raised by the increasing positioning of corporations as citizens in their own right. Not everyone is convinced that the good corporate citizen is possible. After all, corporations are business entities established for the purpose of turning a profit for their shareholders. Some critics accept the portrait of corporations as natural persons, but only to paint them as a particular type of natural person, namely, psychopaths (Figure 6.1). The diagnosis is based on the notion that, like psychopaths, corporations display the following antisocial traits: callous disregard for the feelings of other people, the incapacity to maintain human relationships, reckless disregard for the safety of others, deceitfulness (continual lying to deceive for profit), the incapacity to experience guilt and the failure to conform to social norms and respect the law (Bakan 2004). Others claim that corporations, as artificial entities, are intrinsically *amoral* rather than immoral entities (Stephens 2002). In any case, regardless of whether one sees them as moral, immoral or amoral entities, there is no question that corporations are powerful political agents.

Much of the anxiety about corporations derives from the fact that they are exceptionally powerful citizens, so much so that their power and the way they wield it make them more like governments than individual citizens. This association between corporations and governments goes back a long way. The progenitors of the modern corporation were government chartered corporations, often established to lead colonial ventures, such as the Dutch East India Company (established in 1602), and the most famous of all, the British East India Company (established in 1600). Following the rise of classical liberalism and *laissez-faire* economic theory, corporations transitioned from being government entities to being economic entities free of direct government intervention. Over time, this transition has seen the flow of influence go in the opposite direction. If initially governments influenced business corporations, modern corporations have become adept at influencing governments, to promote their interests and protect their investments. This influence has increased significantly in recent times.

The process of neoliberal globalization has enabled the emergence of massive transnational corporations (TNCs). Some of the largest world economies are TNCs whose revenues are often larger than the gross domestic product (GDP) of many countries, especially those from the Global South. In 2010, Walmart's revenue was larger than the GDP of over 150 countries, including Norway's (the 25th largest economy). There are conceptual problems with such comparisons, because corporate

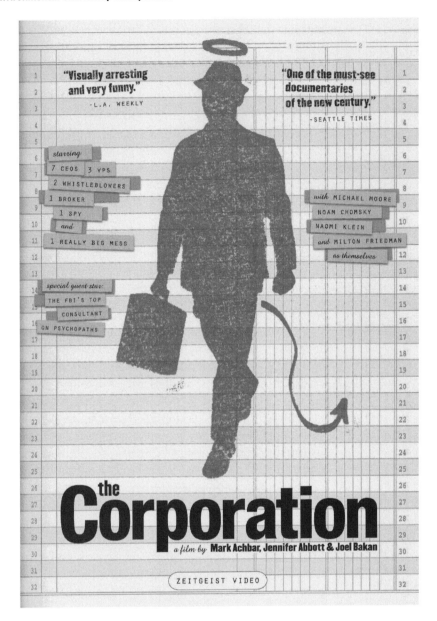

Figure 6.1 The Corporation *(2003)*

Source: http://thecorporation.com/

revenue is not equivalent to GDP, which is measured in terms of added value. But even when one accounts for the lack of strict equivalence, the fact remains that, in 2002, 29 of the world's 100 largest economies were companies (Roach 2007). Their massive scale and transnational reach have made these corporations powerful global political agents. Some commentators have gone as far as to argue that 'the transnational corporation has virtually supplanted the nation state as the central institution dominating the lives of people in most parts of the world' (Clarke 1999: 158).

Corporations have typically used their power to influence governments in economic matters, e.g. to deregulate financial markets and capital flows, reduce corporate taxes, minimize labour regulations and resist environmental regulations. This influence from outside has been increasingly complemented by participation from inside, through their presence in network-based governance. As we have seen in Chapter 5, the shift from government to governance has enabled corporations to move beyond influencing government to being actively involved in the actual process of governing. This makes corporations operate more like actual governments. Indeed: 'A social actor with the resources and power of a large multinational that *participates* in governance becomes increasingly indiscernible from the government as the ultimate and sovereign *authority* over these governance processes' (Crane *et al.* 2008a: 51, original emphasis). Some authors have even concluded that 'corporations rule the world' (Korten 1995). In any case, irrespective of our views on who rules the world, corporations have enormous political power, giving good reason to view them not only as citizens (or quasi citizens) but also, and perhaps more so, as governments (or quasi governments) (Crane *et al.* 2008a).

Corporate environmental citizenship (CEC)

The corporate sector is often accused of being largely responsible for some of the most severe social and environmental problems, including social inequality, global poverty, resource depletion, pollution and deforestation. Until recently, most companies saw social and environmental issues (e.g. labour conditions and waste management) as inhibitors on profitability and beyond their actual responsibilities. These responsibilities were narrowly defined in terms of the maximization of profits for their shareholders, or as Milton Friedman famously put it: 'There is one and only one social responsibility of business – to use its resources and engage in activities designed to increase its profits' (1970). In this context, it is hardly surprising to find out that, for a long time, corporations did very little to mitigate their impact on the environment beyond compliance with environmental legislation. Thus, for example, their general attitude to waste management came to be associated with the acronym CATNAP ('cheapest available technology, narrowly avoiding prosecution') (Smith and Pangsapa 2008: 149). This minimalist approach has been replaced with a more proactive engagement with social and environmental concerns since the 1980s. This section outlines this change of attitude and examines the main debates and questions raised by the corporate embrace of environmental citizenship.

A brief history: from CSR to CEC

Over the course of the last 30 years or so, the legal obligation of business corporations to operate for profit has become complemented by the increasing prevalence of the notion of corporate social responsibility (CSR). This concept encourages a new business paradigm in which companies have broader responsibilities than simply making profits for their shareholders. The new paradigm replaces the traditional single bottom line (economic profits) with the now famous 'triple bottom line': economic profits, social accountability (e.g. human rights and labour conditions) and environmental sustainability. This approach has also come to be known as the three Ps: Profits, People and Planet.

CSR came to prominence in the 1990s and has since been grafted onto the concept of corporate citizenship, a buzzword that emerged during the 1990s and has gained prominence in the 2000s. Since then, corporations have made it a part of their identity to produce and publish 'corporate citizenship' reports, actively publicizing their status as good corporate citizens. Irrespective of how those reports are received by the general public, they illustrate the increasingly complex public identity of corporations. The most accomplished reports provide excellent illustrations of the corporate citizen in all its facets, reflecting a multidimensional conception of corporate citizenship that evokes the best portraits of good active citizenship. See, for example, the annual reports produced by Disney and Walmart. These and plenty of other reports also illustrate how the concept of sustainability has come to dominate much of the discourse of CSR, giving birth to the concept of CEC.

CSR is now a career and an industry in its own right, with full-time staff, websites, newsletters, professional associations and armies of consultants.

The concept is so mainstream that it has amongst its foremost supporters the World Business Council for Sustainable Development. Created in 1995 and based in Geneva in Switzerland, the Council has a membership of over 200 corporations, including British Petroleum, Coca-Cola, Dow Chemical, DuPont, Ford, General Motors, Nestlé, Procter & Gamble, Shell, Sony and Time Warner. Their motto is 'business solutions for a sustainable world'. Their website states that 'membership is open to companies committed to sustainable development and to promoting the role of eco-efficiency, innovation and corporate social responsibility'.

The emergence of climate change as the dominant environmental theme in the public imagination has positioned environmental sustainability at the heart of CSR. The social acceptance of sustainability is such that companies can no longer ignore the issue. The significance of this pressure was perfectly captured by UN Secretary-General Ban Ki-moon, who, in one of his op-eds, made the point that business as usual no longer works, and that the private sector needs to understand that twenty-first-century 'corporate responsibility means corporate sustainability' (2012b). In other words, what is at stake is not only the state of the environment but corporate survival itself. Corporations have responded by internalizing sustainability, particularly by incorporating sustainability reporting into their organizational strategies (Cintra and Carter 2012).

CEC can be seen as a different way to refer to CSR or as a mere offshoot of CSR. However, there are two important points of differentiation. First, CEC makes explicit the ascendancy of environmental issues within the discourse of CSR. Second, the reference to citizenship reinforces the notion of corporations as citizens and positions them at the centre of discussions regarding the meaning and practice of citizenship in general and environmental citizenship in particular. In other words, the shift in terminology suggests that environmental citizenship is not something that happens to corporations but is an arena where they have become central actors. In practice, CEC has come to designate the activities and organizational processes adopted by businesses to meet their environmental responsibilities. These initiatives are many and varied, and are typically publicized in their corporate citizenship and sustainability reports (see Box 6.1).

The most significant way in which businesses have embraced environmental concerns is through their widespread adoption of the concept of sustainable development. This concept suggests that economic development and environmental protection are compatible, even

Box 6.1

Corporate adoption of environmental citizenship

This study identifies and classifies the practices of multinational corporations (MNCs) in the field of environmental citizenship and their contribution to sustainable development. The findings are the product of the content analysis of 38 environmental performance reports from a wide range of MNCs, including: Amoco, British Petroleum, Chevron, Ford, General Motors, Monsanto, PepsiCo, Shell, Sony, Unilever, Texaco and Toyota.

Externally-oriented corporate citizenship programmes

1. *Incentives for employee collaboration with stakeholders for environmental improvement*: provide incentives and encouragement for corporate employees and managers to collaborate with external stakeholders and local communities to improve environmental conditions and prevent or remediate environmental degradation.
2. *Philanthropic support for environmental activities*: support philanthropic activities in communities where facilities are located and in support of environmental groups dealing with worldwide environmental problems (e.g. wildlife habitats and biodiversity).
3. *Strategic alliances with environmental groups and communities*: develop strategic alliances with environmental and public interest groups to solve environmental problems.

Internally-oriented social responsibility practices

1. *Enhanced regulatory compliance*: adopt voluntary practices to reduce negative environmental impacts of hazardous emissions in the communities where the companies are located, beyond what local or national regulations require.
2. *Pollution prevention and clean manufacturing practices*: adoption of clean manufacturing practices that prevent pollution before it occurs, not only in their own operations but also amongst their suppliers, vendors, and contractors.
3. *Product and process redesign*: redesign products and processes to reduce the environmental impacts of products over their lifetime, and achieve more beneficial environmental impacts for customers and communities.
4. *Materials reduction, recycling, and re-use*: reduce waste materials or re-use them instead of disposing of them at or outside of plants and facilities.
5. *Resource conservation*: reduce the use of energy, water and other natural resources.

The study reveals that externally-oriented interactions with stakeholders and communities, although visible and important, account for only a small percentage of the environmental expenditures. Most resources are committed to internal management practices that both generate financial returns for the company and produce beneficial environmental results.

(Rondinelli and Berry 2000)

complementary. In other words, sustainable development – particularly as deployed by the business sector – suggests that investment and growth can be reconciled with environmental concerns. However, the continuing centrality of the traditional bottom line (i.e. economic profits) has led critics to argue that the language of responsibility and sustainability is merely a stratagem or tool of corporate packaging to counter criticism of the corporate sector's environmental record. Timothy Doyle and Doug McEachern, for example, note that business commitment to sustainable development 'has not meant great policy innovation, but when business needs to present a case for environmental concern then it can be packaged using this concept' (2008: 208). In so far as CEC has become part of the corporate packaging to wrap environmental criticism, similar questions apply: is CEC little more than business as usual? Has CEC in fact opened the door to the latest marketing dream, namely greenwashing? The answers to these questions, explored in the next section, will shape the long-term impact of CEC.

CEC: business as usual and/or greenwashing?

The enthusiastic corporate embrace of environmental citizenship has been driven by the recognition that CEC is 'good for business'. Although CEC is always framed in terms of a win-win, many studies suggest that profit is invariably the decisive factor in adopting any kind of environmental initiative. The analysis conducted by Rondinelli and Berry (2000) of the environmental reports of 38 MNCs shows that the main initiatives adopted by corporations relate to internal management of materials, products and processes expected to generate financial returns. Theirs and other studies indicate that 'inasmuch as corporate citizenship may be desirable for society as a whole, it is unlikely to be widely embraced by organizations unless it yields concrete business benefits' (Maignan and Ferrell 2001: 479).

Numerous other studies have illustrated the connection between 'going green' and a range of specific benefits, including: cost savings, enhanced revenue, improved efficiencies, increased productivity, risk mitigation, competitive advantages, business opportunities and enhanced brand image (e.g. Prakash 2000). In addition, the benefits of a strong reputation for good corporate citizenship can include greater access to capital, increased customer loyalty and 'less regulatory oversight or more favorable treatment by regulatory agencies' (Rondinelli and Berry 2000: 74). These studies support the notion that corporations are ruled by 'the tyranny of the bottom line' (Estes 1996). Furthermore, given the emphasis on self-regulation, corporate responsibility ultimately 'depends on the willingness of companies to develop robust codes of conduct and monitor their own activities and those of subsidiaries, as well as contracted outsourced manufacturers and suppliers' (Smith and Pangsapa 2008: 158).

The extent to which these codes and practices can result in significant environmental improvements, particularly in the long term, remains a matter of contention. The debate revolves largely around the actual purpose and the genuineness of the corporate embrace of environmental citizenship. In fact, much recent scholarship casts considerable doubt on the ability of CSR (and by extension CEC) to make tangible progress (e.g. Vogel 2005) and identifies instances of corporate 'greenwashing', where companies put forward an environmentally friendly image whilst doing little to reform their practices (e.g. Karliner 1997; Gillespie 2008).

The term 'greenwash' entered the *Oxford English Dictionary* in 1999. The dictionary defines it as: 'disinformation disseminated by an organization so as to present an environmentally responsible public image'. Given that the creation of a green image is a useful mechanism to increase profits, perhaps it should not be surprising that greenwashing is widespread, particularly in advertising (Coppolecchia 2010). Business and marketing scholars explore (and often advocate) the need to create and/or enhance an environmental image as part of the business strategy (e.g. Kashmanian *et al.* 2011). Indeed, corporate citizenship has been openly described as a 'marketing instrument' (Maignan and Ferrell 2001). This image enhancement is prevalent on company websites, which are increasingly adopting the language of environmental citizenship. The authors of one recent study about the communication of CEC on Fortune 500 websites even recommends companies place 'a single "Environmental Citizenship" link prominently on the home page' (Yu *et al.* 2011: 205).

The advent of social media promises to add another dimension to the battle over corporate image. These are still very early days, but the debate is already set between those who see social media as yet another tool for corporate greenwashing and those who see potential for democratic accountability in this new way of relating between corporations and citizens. Early studies suggest that social media will likely diminish the incidence of corporate greenwashing, by helping to expose it and thus creating a culture of transparency where corporations will find it harder

Figure 6.2 *'Corporate greenwashing'*

Source: Cartoon courtesy of Cathy Wilcox.

(and increasingly costly) to hide their bad corporate behaviour (Lyon and Montgomery 2013). Others go further, and argue that there is a strategic fit between social media and corporate responsibility, by which social media rewards the socially responsible firms and punishes the socially irresponsible ones (Lee *et al.* 2013).

There are several corporate practices – in addition to greenwashing – that cast serious doubts over corporate green credentials. Some of the major corporations (especially in the energy sector) continue to fund projects and think tanks that challenge the scientific consensus regarding climate change (e.g. the American Enterprise Institute). They also fund initiatives such as the 'wise-use movement' and so-called 'astroturf campaigns' (i.e. fake grassroots political campaigns) aimed at counteracting the environmental movement (Beder 2002). The continuing poor environmental record of corporations in the Global South (especially energy companies) is also a matter of public record (Maiangwa and Agbiboa 2013). In addition, the fact that many green brands are being snapped up by large corporations adds to the suspicion that these are only interested in profits, and are even willing to compromise genuine green business models (i.e. companies that have made sustainability central to their operations) in order to increase their profits, and add a 'green tinge' to their image along the way (MacDonald 2011).

The extent to which the corporate green turn has become an exercise in marketing is perhaps best symbolized by their takeover of Earth Day in the United States. As Leslie Kaufman (2010) reported for *The New York Times*: 'At 40, Earth Day Is Now Big Business'. The article captures perfectly the contrast between the celebrations on the first Earth Day in 1970 and those of 2010. In 1970, she writes, 'organizers took no money from corporations' and 'held teach-ins to challenge corporate and government leaders' (Kaufman 2010). Forty years later, she notes, 'the day has turned into a premier marketing platform for selling a variety of goods and services, like office products, Greek yogurt and eco-dentistry' (Kaufman 2010). She also ironically notes how: 'F.A.O. Schwartz is taking advantage of Earth Day to showcase Peat the Penguin, an emerald-tinted plush toy that, as part of Greenzys line, is made of soy fibers and teaches green lessons to children' (Kaufman 2010). Rebecca Tarbotton expressed similar sentiments in an article for the *Huff Post Green* entitled 'Has Earth Day Become Corporate Greenwash Day?' She writes:

> I can green my dog's food, I can green my breath freshener, and I can even have green coal and oil companies (if you believe the sponsors of some of this year's Earth Day events). . . . Corporations from F.A.O. Schwartz to Chevron have become very good at hyping their company's green image, and making Earth Day their day.
>
> (Tarbotton 2010)

The corporate embrace of environmental citizenship will continue to be a matter of much debate between those who see it as a step in the right direction and those who view it as a move that masks the true nature and depth of the environmental problems the world is facing, and which, in their view, cannot be solved by profit-driven corporations. But, in any case, and irrespective of whether their green turn is genuine or merely strategic, business as usual and/or greenwashing, no one can doubt the increasing influence that corporations have on the current articulations of environmental citizenship.

CEC: globalizing neoliberal environmental citizenship

Corporations have long influenced the meaning and practice of citizenship, but their explicit incorporation of environmental citizenship into their identities and practices has deepened that influence in recent times. In this section we will explore the impact that corporations in general, and their green turn, have on the three constitutive elements of citizenship: membership, rights and duties. That impact, as we shall see, translates into the promotion of a specific notion of environmental citizenship, namely: neoliberal environmental citizenship.

The ascendancy of the global corporate citizen

In general terms, corporations contribute to the promotion of global conceptions of citizenship. This arises, first and foremost, from their embrace of globalization, which, by definition, raises the profile and prospects of global citizenship. The fact that corporations adopt the language of 'the global' in many of their initiatives (e.g. the Global Compact) and support global/ist projects (e.g. the Earth Charter) also promotes the sense that we are all members of one global community. Corporations also advance the prospects of cosmopolitan citizenship through the challenge they pose to national sovereignty. Importantly, corporations do not advocate the dissolution of national borders, let alone of nation-states. After all, they rely on states to maintain stability and create conditions that are friendly to property and to the free operation of markets. Corporations prefer a system where the state retains the capacity and authority to control its physical borders, regulate the movement of people, and prevent large population dislocations that can upset a predictable business environment. However, they do advocate the opening of borders to business and capital, to facilitate the free flow of goods, services and capital. This enables an increasing number of people to access products from all corners of the globe, expanding the potential for interconnection (if not identification) with people from around the world; but it also displays the globalization-à-la-carte promoted by corporations – one that privileges the interests of the corporate citizen, often at the expense of individual citizens.

The behaviour of corporations and their executives reveals some significant tensions between global and national identities. On the one hand, the multinational and transnational identities and operations of TNCs invoke notions of global citizenship. However, corporations often play with different global and national identities for marketing purposes, depending on what type of identity might be more profitable for a particular product or campaign. Moreover, despite their global presence, corporate identity is still often tied to particular nations. General Motors and Coca-Cola, for example, might be transnational corporations but they remain quintessentially American. In addition, corporations have their legal headquarters in a nation-state, most of them in the Global North. Similar to natural citizens, their legal home is typically their nation of origin, but sometimes they relocate all or part of their headquarters (e.g. their legal and/or financial residence) to other nations, typically for financial reasons (e.g. Apple in Ireland). Something similar occurs with their frequent-flyer corporate executives. Like the companies they manage, executives are also formal members of a nation-state, typically that of their birth (their original nationality), but often adopt a second or third residence (or nationality) for business or taxation purposes. Their lives, however, separate them, physically, emotionally, psychologically, and in terms of their identity politics, from any of their fellow nationals. These members of the international business elite often have more in common with each other than with average citizens with whom they happen to share the same nationality (Freeland 2012). In short, corporations generate for themselves and their affluent frequent-flyer managers a sense of global identity that contributes towards a sense of global citizenship, but one that is largely restricted to the corporate brand and the managerial elite.

The corporate take on global citizenship obscures the profound inequalities amongst citizens and nations that corporations often exploit for profit, such as when they relocate production to countries with lower productions costs and weaker environmental protections. Indeed, corporations tend to exploit both cultural and material differences: as products to be sold (cultural differences), in the first case; and as weaknesses to be exploited (wealth differences), in the second. The double standards of corporate behaviour regarding their operations in the Global North and the Global South are well documented, not least by the activist collectives dedicated to researching their actions, such as Corporate Watch. These double standards include corporations treating the citizens and nations of the Global North as first-class and those of the Global South as second-class. These double standards, and their tragic consequences for the citizens and activists of the Global South, sometimes come to the fore in a very public fashion, as was the case in Nigeria during the 1990s, when several Ogoni activists, including

Ken Saro-Wiwa, were executed before the passive eyes of the company responsible for much of the damage they were campaigning against, namely, the oil company Shell. Reflecting on citizenship questions and the environmental crisis in the Niger Delta in the wake of the Ogoni crisis of 1995, Wunmi William (2002) notes how the same company, which operates according to laws in the extraction of resources in Europe, neglects to adhere to laws in most West African countries.

In short, despite all the corporate rhetoric of global citizenship and references to the notion that we are all in this together, as members of a single global community, the actions of corporations betray a far from egalitarian outlook and disposition. To paraphrase George Orwell: 'all citizens are globalized, but some are more global than others'. In particular, corporations (corporate citizens) are more global than humans (natural citizens), and corporate executives are more global that the vast majority of world citizens. In addition, corporations treat citizens and nations of the Global North and the Global South as first- and second-class groups, respectively. The inequality masked by corporate notions of global citizenship becomes more evident when we look at the nature of rights and duties promoted by notions of corporate citizenship in general and CEC in particular.

The ascendancy of property rights

Locke's notion that 'government has no other end, but the preservation of property' is one which corporations could not embrace more wholeheartedly. Since their inception, corporations have been key agents in the promotion of property rights and the exporting of liberal citizenship around the world (Crane *et al.* 2008b). Indeed, the centrality of property rights in their articulation of citizenship makes corporations the quintessential liberal citizen. In recent times, their influence has positioned property rights at the heart of citizenship in general and environmental citizenship in particular. The ascendance of property rights in conceptions of environmental citizenship has important implications for two interrelated issues: the articulation of the relations between humans and nature; and the articulation of the relations between corporations and indigenous peoples. The former manifest themselves in the treatment of nature as capital. The latter manifest themselves most notably in conflicts about mining and extraction of resources and the assignation of property rights to traditional knowledge (Crane *et al.* 2008b).

Corporations have always promoted a conception of nature as capital and tried to exert property rights over natural resources. The concept of 'natural capital' has been extended in recent times to include 'ecological services', that is, services provided by nature (e.g. water purification by wetlands, crop pollination by bees, air purification by trees). Supporters of attaching a price to ecological services (also known as 'ecosystem services') argue that this will address some of the negative externalities (i.e. environmental costs not taken into account when pricing goods and services) and thus improve the treatment of the environment. For example, putting a price on bee pollination of crops could enhance the care corporations take to prevent the disruption of bee habitats and colonies, in order to avoid having to pay reparations or financial compensation for the damage inflicted. The same rationale informs the pricing of carbon and the creation of tree-planting schemes that treat trees as service providers of carbon sequestration. The treatment of nature as property is also present in the increasingly popular practice of bio-offsetting, that is, compensating for biodiversity losses related to a particular project with improvement to biodiversity protection somewhere else. This practice turns nature into property by developing a metric through which exchanges of biota can be carried out. Irrespective of the feasibility and environmental merits of such schemes, the incorporation of ecological services and biological resources into the normative framework of property rights extends the reach of neoliberal environmental citizenship – one where everything, including nature, is treated as a commodity to be bought and sold in the marketplace (Figure 6.3).

The ascendancy of property rights comes also at the expense of alternative conceptions of citizenship that reject the concept of nature as property, particularly those held by indigenous peoples. There is no uniform indigenous conception of citizenship, environmental or otherwise. However, historically and by and large, indigenous peoples around the world have not characterized their relationship to the land in terms of private property rights. Instead, territories and their resources are typically held and managed as a commons, and the relation to the land is based on rituals and cosmologies that view humans as part (and custodians) of the land. This picture has been radically altered with the arrival of corporations dedicated to the exploitation of natural resources, a process that began during the colonial expansion of Europe but has accelerated since the twentieth century. Leaving aside the exploitative nature of the colonial (and postcolonial) corporate enterprise on indigenous lands, the deployment of property rights as the formal

Figure 6.3 *'Ecosystem services as Trojan horse'*

Source: Cartoon courtesy of Cynical Conservationist.

normative code to engage with the land and its resources undermines the emergence of alternative conceptions of environmental citizenship based on indigenous cosmologies.

The challenge to indigenous conceptions of citizenship has intensified in recent times, most notably through the application of property rights to traditional systems of knowledge. For example, corporations have sought to patent plants and herbs that have been used for medicinal purposes by indigenous peoples in their native lands for hundreds, if not thousands, of years. With the active collaboration of governments, who must sign and enforce the relevant agreements, corporations have begun to impose the agenda of intellectual property rights onto indigenous knowledge through international trade organizations and agreements such as the Agreement on Trade Related Aspects of Intellectual Property Rights (TRIPS) (Banerjee 2008). Critics of this approach refer to the TRIPS regime as 'biopiracy' (Shiva 1997; Bautista 2007). Some have sought a solution by calling for the extension of intellectual property rights protection to traditional knowledge. Whilst this might serve as a quick fix to the continued expropriation and exploitation of indigenous intellectual common goods by private corporations, it would also have some serious implications. First, it would consolidate and entrench the neoliberal private property-based model of environmental citizenship. And second, and more broadly, it would undermine notions of the public good and contribute to the erosion of traditional collectivist notions of citizenship.

The ascendancy of voluntary duties

The fundamental responsibilities corporations owe are to their shareholders. However, corporate citizenship in general and CEC in particular mean that responsibilities have been extended to employees, customers, suppliers and regulators, as well as the broader society in which firms operate, and even future generations. The main duties that have come to be formally associated with global CEC are specified in the world's largest corporate citizenship and sustainability initiative, the UN Global Compact. The compact represents the culmination of earlier efforts to bring together private corporations, national governments and NGOs to encourage self-regulating corporate responsibility and better environmental performance (Smith and Pangsapa 2008: 148). Importantly, the compact is not a regulatory instrument but 'a value-based code of conduct to promote institutional change' (Wagner 2004: 285). The scheme is the equivalent of a public pledge by which corporations promise to support and abide by ten principles, including three environmental principles: supporting a precautionary approach to environmental challenges; undertaking initiatives to promote greater environmental responsibility; and encouraging the development and diffusion of environmentally friendly technologies (see Box 6.2).

The environment principles are firmly located within the paradigm of ecological modernization and illustrate the belief in the capacity of business to combine and achieve economic and ecological rationality. This guide for management is meant to support business so that it can engage in practices that are proactive and exceed the current demands of environmental regulations and laws, but it is entirely voluntary. The Global Compact operates under the auspices of the UN, but this has no authority, resources or political will to effectively monitor, verify and enforce its principles. The initiative is designed as a forum to promote learning through discussion, a framework that encourages involvement and contributes to a flexible learning curve. The project operates through and promotes self-regulation and, since its strategic alliance in 2006 with the Global Reporting Initiative, relies on the release of corporate reports to showcase compliance with the principles of the Global Compact. This approach to corporate responsibility contributes to a softening of corporate duties, turning them into voluntary commitments with a broad pedagogical value.

The only duty as such – and not even this is enforceable – is to monitor and report, to compile data and share tips, ideas and information with

Box 6.2

The ten principles of the UN Global Compact

The UN Global Compact is a set of ten principles in the areas of human rights, labour standards and environmental practices that companies are encouraged to follow and promote. The compact resulted from the joining of efforts of the United Nations and the International Chamber of Commerce (ICC), and was officially launched on 26 July 2000. Since then the consortium has grown to more than 10,000 participants, including over 7,000 businesses in 145 countries. The principles derive from the Universal Declaration of Human Rights, the International Labour Organization's Declaration on Fundamental Principles and Rights at Work, the Rio Declaration on Environment and Development, and the UN Convention Against Corruption.

Human rights

Principle 1: Businesses should support and respect the protection of internationally proclaimed human rights; and
Principle 2: Make sure that they are not complicit in human rights abuses.

Labour

Principle 3: Businesses should uphold the freedom of association and the effective recognition of the right to collective bargaining;
Principle 4: The elimination of all forms of forced and compulsory labour;
Principle 5: The effective abolition of child labour; and
Principle 6: The elimination of discrimination in respect of employment and occupation.

Environment

Principle 7: Businesses are asked to support a precautionary approach to environmental challenges;
Principle 8: Undertake initiatives to promote greater environmental responsibility; and
Principle 9: Encourage the development and diffusion of environmentally friendly technologies.

Anti-corruption

Principle 10: Businesses should work against corruption in all its forms, including extortion and bribery.

(UN Global Compact: www.unglobalcompact.org/)

others. The logic is that sharing ideas and the associated peer pressure that comes with exposure to the best practice of others will generate change. In addition, reporting and transparency will facilitate the creation of informed consumers. As people and corporations learn about environmental issues and their own environmental impacts, they will gradually embrace the duty of caring for the environment, altering their behaviour through self-regulation. Governments are only required to assist with the process of learning (e.g. media campaigns and educational programmes). The decisions are made by corporations and consumers in the marketplace. Thus, despite much talk of duties and action, this is not a republican conception of environmental citizenship, in which duties are owed to the collective and expressed in terms of the public good, but a neoliberal one in which duties are apolitical, voluntary and carried out by corporations and consumers as individuals acting for their own private benefit.

Supply-chain(ed) citizenship: the outsourcing of duties

The way in which corporations, especially the large consumer-oriented retail ones, enforce environmental standards along their supply-chain produces particular dynamics whose study offers additional insights into the impact of corporate behaviour on environmental citizenship. The practices of these retailers have inspired the concept of supply-chain citizenship, defined as: 'a collective of long-distance promises of care that are economically and politically backed up by transnational corporations' (Partridge 2011: 100). This articulation of corporate citizenship shares the same traits as neoliberal citizenship in terms of identity (limited and limiting forms of global belonging) and rights (the centrality of property rights), but illustrates and emphasizes the outsourcing of responsibilities that is becoming central to corporate citizenship in general and CEC in particular. In this section we will explore this conception of citizenship, with particular attention to its paradigmatic illustration, the retail giant Walmart.

Reducing environmental impact through the control of the supply chain has become increasingly popular amongst companies with an expressed commitment to environmental responsibility. This is typically associated with the notion of 'product lifecycle' and related concepts such as 'zero waste' and 'cradle to cradle' that take sustainability as a matter of design (McDonough and Braungart 2002). One of the first companies to embrace this approach was Fuji-Xerox, a company long praised for its commitment to what they call 'sustainable earth environment'. In 2004, Fuji-Xerox issued a new set of environmental, health and safety requirements, adding to the classic three Rs (Reduce, Reuse, Recycle) a fourth R: Refuse. This signified their refusal to operate with suppliers that did not match their environmental expectations, namely, the elimination of all greenhouse gases from the production process (Smith and Pangsapa 2008). Since then, other companies have embraced the cradle-to-grave approach, amongst them B&Q and BMW.

However, the paradigmatic example of supply-chain citizenship is Walmart. Indeed, through its supply-chain management, Walmart has become the emblematic global enforcer of CEC (see Box 6.3). This giant American corporation has a well-established reputation for deploying its company muscle to compel suppliers to comply with the standards they set. In the eyes of many, the fact that Walmart has 'gone green' represents a massive win for environmental sustainability and a positive reflection of the value of CSR and CEC. Walmart has received widespread praise for its efforts by experts and environmental groups, including high-profile environmentalists such as David Suzuki. In a keynote address given at a Walmart's annual business meeting, Suzuki stated: 'Walmart's commitment to sustainability acts as an inspiration and incentive to other corporations to follow suit. The company has enormous influence on corporate thinking and I am delighted with the priorities it has selected' (cited in Ross 2010: 21).

Walmart's main initiative in greening its supply chain is the establishment of a Sustainable Product Index. Announced in July 2009, the Index is a voluntary scheme which requires its 100,000 plus suppliers to 'provide information on the environmental footprint of their products and practices, which will yield a consumer-friendly decision-making tool in a few years' (Cutting et al. 2011: 297). The objective is to create a transparent and ethical supply chain, from the raw materials all the way through to disposal. The key to the scheme is simple: information. The optimistic assumption is that once information on the lifecycle of products is publicly available, consumers will make greener choices and producers will respond to demand for greener products, and this will culminate in structural change within the retail sector. Critics argue that the scheme might not work as intended. For one thing, it assumes that information is enough to produce important consumer change. But, more importantly, the scheme does not challenge the culture of consumerism that many consider to be at the heart of the current ecological crisis.

Box 6.3

Walmart: the global enforcer – our profit is your duty

The Walmart effect

Walmart is the largest retail corporation in the world, with some 100,000 global suppliers and more than $460 billion in revenue in 2012. The power it exerts over its suppliers is well known. The famous story of how Walmart compelled its suppliers to abandon the practice of putting deodorants in boxes in order to save on costs (with incidental environmental benefits) is legendary. It is this kind of power that inspired the so-called 'Walmart effect' (Fishman 2006). The size and power of Walmart are also what has made their 'green turn' one of the most studied and welcomed developments in corporate global business in recent times. Their green goals – as stated on their website – are: to be supplied 100 per cent by renewables; to create zero waste; and to sell products that sustain people and the environment. This is to be achieved through monitoring their own energy consumption and moving towards the creation of a sustainability index for its suppliers. Optimists hope the Walmart effect will green global retail. Critics and sceptics point to Walmart's business model.

The business model

Walmart is renowned for being a 'cost-squeezer'. Its business model has always been one of relentless slashing of prices, undercutting price competition and offering a vast selection of products – all of which contributes to a culture of consumerism that is viewed by many as the antithesis of sustainability. The green credentials of Walmart are also undermined by other core practices of its business model that enable such low prices, in particular, the situation of its businesses in gigantic warehouses on the outskirts of urban areas, and the large-scale distant sourcing of produce. In addition, critics note that the green initiatives adopted in its warehouses, such as recycling cardboard, dimming lights, and installing smaller flushers on toilets, are more about their bottom line than about any sense of corporate environmental responsibility. They ask: would Walmart take up green initiatives that would reduce profits or incur financial costs? The answer seems to be a clear 'no'. Recent reports published in *The Atlantic* and *Mother Jones* indicate that Walmart has pulled back from its sustainability programmes – except for the ones that save money (Schell 2011; Kroll 2012). Yet, Walmart continues to enforce environmental citizenship onto others, mainly its suppliers. The citizenship model operating seems to be: Your Duty, Our Profit.

Resources

Reclaim Democracy: www.reclaimdemocracy.org/walmart_links/
Walmart Community: www.walmartcommunity.com/
Walmart Corporate: http://corporate.walmart.com/

The main critique of supply-chain citizenship is the fact that this approach effectively outsources corporate responsibilities to the suppliers, who find themselves chained to standards of production they cannot challenge, even if they consider them unfair. In the current global economic context, this effectively means that corporations (from the Global North) transfer their own environmental responsibilities (and those of consumers in the Global North) to the producers and suppliers in the Global South. Thus, for example, by pushing the costs onto the suppliers, Walmart discharges it environmental duties by simply transferring the environmental costs of its products from its consumers (largely located in the Global North) to its producers and suppliers (largely located in the Global South). This model of citizenship constitutes an unethical imposition of the North on the South that neatly sidesteps the capacity of governments and citizens of the Global South to control the environmental consequences of the scale of consumption by consumers from the Global North (Figure 6.4). The result is arguably a form of (neocolonial) governmentality that compels local suppliers and firms to internalize global standards set principally in the affluent nations of the Global North (Partridge 2011).

This approach to environmental citizenship also represents a shift of corporate environmental responsibilities onto the consumers. In fact, the consumer is typically presented as the ultimate authority, the sovereign. The corporate approach shifts the emphasis from the consumer's 'right to know' to the consumer's 'duty to know', highlighting the duty to exercise responsible choices when it comes to purchasing products.

WHY CHINA'S CARBON FOOTPRINT IS SO LARGE

Figure 6.4 *'China's carbon footprint'*

Source: Cartoon courtesy of Chris Madden.

The corporation is simply presented as the enforcer of a supply-chain regime determined by the choices of the consumer. Indeed, Walmart and other retail corporations that have embraced the 'green turn', such as Whole Foods Market (Johnston 2008), often offer both 'green' and 'grey' products, thus making sure they capture both the 'green' and 'grey' customer. This is a different way of articulating a 'win-win' business model: whether you buy grey or you buy green, the corporation always wins. Thus, supply-chain citizenship effectively empties CEC of corporate responsibilities.

The model of consumer and supplier citizenship promoted by this supply-chain scheme is one highly exclusionary articulation of neoliberal environmental citizenship: it is a highly individualistic model, completely at odds with other models of environmental citizenship; it is materialistic, thus excluding indigenous ways of being green; and it is completely devoid of any notions of social justice, so has no socially politically or economically transformative potential. It is therefore not surprising that whilst Walmart touts itself on its website as being an active and generous citizen – supporting local events and acting as a prominent benefactor of local communities – it also attracts criticism for the low pay it gives its workers. Walmart has the power to represent itself as the ultimate good (and green) citizen, but does not provide the conditions for its own employees to enjoy substantive citizenship.

Finally, supply-chain citizenship is also effectively a way of displacing government from the sphere of citizenship, replacing government policies and protections with corporate schemes. One can even argue that the power of Walmart relative to its suppliers makes its preferences the equivalent of laws, and Walmart the equivalent of a government. Suppliers ignore the commercial power and authority of Walmart at their own peril. They are effectively subjects of what might best be referred to as supply-chained citizenship. In this context, and as the state regulates less and less and voluntary codes of conduct become the norm, some view NGOs as a counterforce to corporate power that can protect citizens and the environment. However, neither corporations nor NGOs are democratic entities. This has led others to call for more 'supply-chain democracy – that is, political, social, and economic accountability – particularly from the perspective of those for whom ethical standards are being created without their input and without significant improvement of their lives' (Partridge 2011: 107). In the meantime, and in the absence of such democratic protections, supply-chain citizenship represents yet another step in the global ascendancy of neoliberal environmental citizenship.

Summary points

- Corporations have joined citizens and governments as a third agent in the citizenship relationship, but one that can operate variably as citizen and government, depending on the context.
- The power of transnational corporations is posing profound challenges to citizenship in general and environmental citizenship in particular. In the case of environmental citizenship, this has been associated with their adoption of the concepts of CSR and CEC.
- Corporations have embraced the concept of environmental citizenship and have adopted a range of measures to reduce their environmental impact – albeit invariably linked to their financial returns. In some cases, corporate environmental citizenship is mere greenwashing.
- The corporate embrace of environmental citizenship has translated into the ascendancy of global citizenship, property rights and voluntary duties within this conception of citizenship.
- Corporations have driven environmental citizenship towards a neoliberal model, characterized by the emphasis on individual responsibility, voluntary actions and self-regulation. These traits apply both to corporate citizens and consumer-citizens.
- Giant retail corporations are shaping environmental citizenship in ways that have inspired the notion of supply-chain citizenship. This form of citizenship refers to the outsourcing of corporate environmental responsibility to suppliers, and represents a general outsourcing of environmental responsibility from the Global North to the Global South.

Selected readings

Banerjee, S. B. (2008) 'Corporate Social Responsibility: The Good, the Bad and the Ugly'. *Critical Sociology*, 34(1): 51–79.
Beder, S. (2002) *Global Spin: The Corporate Assault on Environmentalism*. Revised edition. Carlton North, Vic: Scribe Publications.
Crane, A., D. Matten and J. Moon (2008) *Corporations and Citizenship*. Cambridge: Cambridge University Press.
Johnston, J. (2008) 'The Citizen-Consumer Hybrid: Ideological Tensions and the Case of Whole Foods Market', *Theory and Society*, 37(3): 229–270.
Karliner, J. (1997) *The Corporate Planet: Ecology and Politics in the Age of Globalization*. San Francisco, CA: Sierra Club Books.
Palacios, J. J. (2004) 'Corporate Citizenship and Social Responsibility in a Globalized World'. *Citizenship Studies*, 8(4): 383–402.
Rondinelli, D. A. and M. A. Berry (2000) 'Environmental Citizenship

in Multinational Corporations: Social Responsibility and Sustainable Development'. *European Management Journal*, 18(1): 70–84.

Yu, J., K. R. Coulson, J. X. Zhou, H. J. Wen and Q. Zhao (2011) 'Communicating Corporate Environmental Citizenship: An Examination of Fortune 500 Web Sites'. *Journal of Internet Commerce*, 10(3): 193–207.

Online resources

The Corporation: The Pathological Pursuit of Profit and Power (2003). Dir: Mark Achbar, Jennifer Abbott and Joel Bakan (uploaded 20 October 2011). Duration 137 mins. Available at: www.youtube.com/watch?v=xHrhqtY2khc. The documentary is also available here: http://thecorporation.com/

Excerpt: 'The Pathology of Commerce' (uploaded 24 June 2007). Duration 1 min. Dr. Robert Hare applies 'the psychopath diagnostic test' to corporations. Available at: www.youtube.com/watch?v=s5hEiANG4Uk&feature=relmfu

Debating Corporate Social Responsibility in End Times: Between Critique and Change (published on 19 March 2013). Duration 106 mins. Available at: www.youtube.com/watch?v=XRMYQLhReZQ

Deutsche Bank: Corporate Citizenship (published on 11 October 2012). Duration 1.53 mins. Promotional video presenting Deutsche Bank as a good corporate citizen. Available at: www.youtube.com/watch?v=YPPFjoeykmM

What Is Corporate Social Responsibility (CSR)? (published on 30 September 2012). Duration 11 mins. Available at: www.youtube.com/watch?v=E0NkGtNU_9w

Why HP: Corporate Citizenship and Environmental Stewardship (uploaded 2 July 2010). Duration 2.15 mins. Promotional video presenting HP as a good corporate environmental citizen. Available at: www.youtube.com/watch?v=buNC-VKV-ss

What Would Jesus Buy? (2007) Dir: Rob VanAlkemade (published on 27 June 2012). Duration 91 mins. Join Reverend Billy Talen and the Church of Stop Shopping Gospel Choir on a cross-country mission to save Christmas from the Shopocalypse (the end of humankind from consumerism, over-consumption and the fires of eternal debt). Available at: www.youtube.com/watch?v=mAxuNdtZt7c

Student activity

Explore the citizenship profile of a (green) business or TNC. Select a transnational corporation, or national or local business, and explore their website trying to identify the conception of environmental citizenship with which they operate. Explore some of the materials in their website: mission statement, annual reports, green initiatives, video galleries, corporate reports, etc.

Discuss the relation of that content to the basic elements of citizenship: membership, rights and duties (individual and collective). What kinds of rights are defended, and for whom? What kinds of duties are defended, and for whom? Does the organization reflect moderate or radical conceptions of environmental citizenship? Is its perspective anthropocentric or ecocentric? Do you think the organization you have chosen reflects a particular conception of environmental citizenship, e.g. liberal, republican, cosmopolitan, ecological and/or neoliberal?

Discussion questions

- Do you think corporations are more like citizens or more like governments? In what ways?
- Do you think CEC is business as usual and/or a mere exercise in corporate greenwashing?
- Do you think 'ecological services' should be priced? Does that solve the problem of negative externalities?
- Is consumerism compatible with green citizenship? Or is it consuming citizenship?

Part III

Pedagogies and representations

7 **Learning environmental citizenship**

> In the end, we will conserve only what we love; we will love only what
> we understand; and we will understand only what we are taught.
>
> Baba Dioum (1968)

Citizens are made, not born. In other words, we learn to be citizens.
To be sure, most people obtain the legal status of citizens at birth, but
that says little about what kind of citizen we will become (e.g. active,
passive, green, etc.). Citizens need some degree of knowledge, skills
and dispositions, as well as capabilities, for their rights and duties to
be meaningful. In fact, who is a citizen and what kind of a citizen one
becomes are a highly manufactured process. The making of citizens,
green or otherwise, is a long and complex process that needs to be
understood in the context of socialization, that is, the learning of norms
and behaviours that allow individuals to become functioning members
of a given society. This chapter explores the main spaces of political
socialization (i.e. schools and the media) and outlines the dominant
articulations of green citizenship that emerge from formal education,
news media and popular culture. The chapter cannot do justice to
the diversity and complexity of green pedagogies so, once again, the
focus will be on its salient aspects and trends. The chapter includes a
whole section dedicated to exploring the most popular and widespread
pedagogical device through which we are being taught to be (and
disciplined into being) green(er) citizens: the ecological footprint.

Green(ing) citizenship through formal education

Education has always been at the heart of citizenship. The best historical
accounts of citizenship leave no doubt that 'education for citizenship is
not an optional extra, but an integral part of the concept' (Heater 2004b:

326). In fact, the model of education has been intimately related to the prevailing model of citizenship at any given time. Moreover, education features prominently in the work of the most influential theorists of citizenship, including Aristotle, Bodin, Locke, Rousseau and Marshall. The significance of education for citizenship was perfectly captured by the US Supreme Court in its 1954 decision on *Brown vs Board of Education of Topeka*. In its famous ruling, the court stated that education 'is the very foundation of good citizenship' (Heater 2004b: 112).

Education is also central to the making of green citizens. Much of what we learn about the environment and how to relate to nature comes from what we learn through formal education, especially in the Global North. This has been particularly the case since the institutionalization of environmental education as part of citizenship education that began in the 1980s. Since then, citizenship education and environmental education have been on a path of convergence, to the point that environmental education is now often referred to as education for environmental citizenship. In this section we will examine the impact of that convergence on the three constitutive elements of citizenship: membership, rights and duties. But we begin with a look at the relation between education, citizenship and the environment.

Education, citizenship and the environment

The origins of environmental education go hand in hand with the origins of the modern environmental movement. Since the early days, environmental activists used the media to campaign on environmental issues and raise awareness and support for alternatives to the practices that were damaging the environment. The educational initiatives of the environmental movement led to the emergence of an international community dedicated to the promotion of environmental education (Postma 2006: 4). One of its early influential figures was William Stapp, who defined the aims of environmental education as 'producing a citizenry that is knowledgeable concerning the biospherical environment and its associated problems, aware of how to help solve these problems, and motivated to work towards their solution' (Stapp *et al.* 1969: 30). Environmental education gained international recognition at the 1972 Stockholm Conference and was officially embraced at the Intergovernmental Conference on Environmental Education, held in Tbilisi in 1977. The Tbilisi Declaration stated that it should be a right of every citizen to receive environmental education, defined as a lifelong learning process designed to increase individual knowledge and awareness about the environment, to develop the necessary skills to address environmental challenges, and to foster values, attitudes and commitments to make informed decisions and take responsible individual and collective action to solve present and future environmental problems (UNESCO 1977).

The formalization of environmental education and its integration into national curricula did not begin until the late 1980s, following the publication of the Brundtland Report. The report recommended the inclusion of environmental education in curricula to foster a sense of environmental responsibility and teach students how to monitor, protect and improve the environment – all this within the framework of sustainable development. The link between environmental education and sustainable development deepened at the 1992 Earth Summit, and was formalized in Chapter 36 of Agenda 21. Since then, governments have integrated environmental education at all levels, stages and aspects of education: schools, colleges, universities, professional and vocational training (Postma 2006). The greening of the curriculum has come to be seen as crucial for promoting sustainable development and improving the capacity of citizens to address environmental issues. The importance of this framework has been reaffirmed by the declaration of 2005–2015 as the United Nations Decade on Education for Sustainable Development.

The inclusion of environmental education in the curricula has reignited the debate about the nature of citizenship education, especially in the liberal democracies of the Global North. The extent to which environmental education should be understood as education *about* the environment (that is, knowledge about the state and functioning of the environment) and/or education *for* the environment (that is, concerned with values, attitudes and positive action for the environment) is the main point of contention. Thus, for example, Derek Bell, coming from a liberal disposition, argues that environmental education 'should provide children with the "mental equipment" and the motivation for informed participation in "sustainability" decisions, but it should not seek to promote specifically "green" ideals' (2004: 48). Bell advocates the compulsory integration of environmental education into the curriculum, but in ways that combine 'education about the environment' (and about some competing environmental ideals, including green, non-green and anti-green ideals) with 'education for sustainability' (2004: 50). In other words, what matters is sustainability, not the 'colour' of sustainability – this should be up for pluralistic public deliberation.

Others, like Gregory Smith and Dilafruz Williams, coming from republican dispositions, argue for greener and more comprehensive models of ecological education. These authors advocate developing a personal affinity with the Earth, engaging in environmental activism, countering the trend towards individualism, considering questions about and fostering social justice, and even questioning the assumptions of modern industrial civilization (Smith and Williams 2000).

The integration of the environment into formal education tends to operate under the assumption that citizens fail to act responsibly only or primarily out of ignorance, and that knowledge will automatically lead to action. This approach, notes Sherilyn MacGregor, suggests that 'it is uneducated and irresponsible individuals – rather than unsustainable and unjust social and economic relationships – who are at the root cause of the environmental crisis' (2006b: 115). But not everyone shares this assumption. Thus, an increasing number of scholars are proposing more sophisticated pedagogical approaches to environmental citizenship. Stephen Gough and William Scott (2006), for example, argue for a model that entails two rounds of learning: the first involves knowledge (the learning of facts); and the second involves collective actions (learning through activities). For their part, Monica Carlsson and Bjarne Jensen (2006) argue for an action-oriented education that includes both the personal and the contextual, that is, knowledge about social and structural changes, not just about individual actions. Others call for locating social justice at the heart of environmental education, making education for environmental sustainability synonymous with education for social sustainability (Hayward 2012).

The greening of citizenship education has translated into a wide array of initiatives, including green lessons, science projects (sometimes in partnership with other agents), class activities (e.g. discussions, video and drama productions) and out-of-class activities integrated into school programmes (e.g. field trips, visits to museums, zoos and aquariums, setting up and running community gardens). Educational institutions have internalized sustainability and increasingly project a green image as part of their public profile, in order to attract students and support. This trend is partly driven by the creation of lists ranking schools, colleges and universities based on how green they are (e.g. People & Planet Green League, in the UK). Indeed, educational institutions have been compelled to embrace environmental citizenship, not only to present themselves as promoters of the concept but also to be seen as good environmental citizens. But what is the impact of environmental citizenship education on conceptions of environmental citizenship? And what kinds of green citizens are being produced by environmental education?

Globalized future-oriented citizens

Environmental education impacts, first and foremost, on our spatial conception of citizenship. In modern times, citizenship education has been explicitly related to the formation of national citizens. Yet, as we have noted numerous times, many environmental issues, most notably climate change, transcend national borders and demand we pay attention to our role as global citizens. Indeed, learning to be a green citizen requires recognizing the ways in which our actions often affect distant others. In other words, environmental education demands we give some consideration to the global political space, i.e. the cosmopolis or, in ecological terms, planet Earth. This creates a significant tension regarding our spatial conception of citizenship. The modern tendency of citizen education to affirm the supremacy of the national space and instil national (even nationalist) sentiments clashes with the global disposition demanded by education for environmental citizenship. In addition, environmental education often invokes the saying 'think globally, act locally', thus presenting the local as the relevant political space for the exercise of environmental citizenship. However, the importance of the national is far from over.

The tension between the global and the national manifests itself in environmental education programmes, both in the Global North and the Global South. Anna Bullen and Mark Whitehead (2005) provide an excellent account of how this tension plays out in Wales. Their analysis suggests that the national frame within which environmental education still operates limits the potential to produce citizens who can project their actions and identities beyond the national political space. The resilience of the national space is more pronounced in countries with strong statist regimes or nationalist identities, such as China. Wing-Wah Law's (2011) work on environmental education in China shows how the state can still play an important role in framing citizenship education in ways that resist global convergence. Indeed, China provides a good illustration of how 'the nation-state continues to be a key actor in nation building, interpreting, responding selectively to growing global influences, defining citizenship, and prescribing citizenship education within its national borders' (Law 2011: 209).

Environmental education also impacts on the temporal dimension of citizenship, particularly promoting its orientation towards the future. To some extent, citizenship education has always been about the future, about transforming the children of today into the citizens of tomorrow. But environmental education projects citizenship beyond the immediate future, bringing into consideration the distant future and the yet-to-be-born generations. Here, we must note that the conception of children as not-quite or future citizens is far from unproblematic. The United Nations Children's Fund (UNICEF) has long been a defender of treating children as active citizens, noting that even though children lack political rights, they take part in political actions, movements and campaigns, as well as in political and armed struggles, not to mention that they provide a vital contribution to social and economic life. In line with this, Bronwyn Hayward makes a compelling case for taking children seriously as political actors, particularly regarding their role in shaping ecological citizenship. She describes how children can learn to be citizens by engaging in democratic decision-making processes in their local environments and argues that we can and must create the conditions for children to become confident, participant ecological citizens, capable of collective imagination and action for a better, sustainable world (Hayward 2012).

There is clear merit in the notion that environmental education should 'create a new generation of citizens who are greener than their parents' (Bell 2004: 43). But we should also consider the possibility that environmental education can and should create greener parents. In other words, forming the adult citizens of tomorrow should not be done at the expense of exempting today's adult citizens from our environmental responsibilities. Otherwise, we run the risk of simply burdening future generations with the environmental responsibilities that we failed to live up to ourselves (Postma 2006). The future-oriented focus of environmental education also runs the risk of sidelining the historical collective responsibility that the industrialized countries of the Global North have for the current environmental predicament, both globally and in relation to the Global South. The risk of ignoring matters of inter-generational and intra-generational justice suggests that environmental education must be comprehensive in its temporal dimension, as well as much more justice-oriented. Indeed, justice demands we take the past, the present and the future into consideration when teaching (and learning) environmental citizenship.

In short, education for environmental citizenship tends to promote the creation of global citizens, but still within a national framework that limits the potential to embrace the global as a space for the exercise of environmental citizenship. This is a globalizing education that produces globalized and glocalized, but not fully rounded global citizens. These are citizens still imagined within national boundaries, and not well attuned to global injustices. This lack of awareness is deepened by the production of future-oriented citizens who are often blind to intra-generational injustices, particularly those derived from historical environmental dynamics. Some of these blind spots are deepened by the dominant portrait of rights and duties associated with currently hegemonic notions of environmental citizenship education, to which we now turn.

Personally responsible (consumer) citizens

Environmental citizenship education shapes the citizens of tomorrow not just in terms of our spatial attachment (local, national, global) and temporal projection (past, present, future), but also in terms of our orientation towards particular conceptions of citizenship as they relate to rights, duties and participation. Joel Westheimer and Joseph Kahne (2004) have identified three main types of citizens produced by different environmental pedagogies: (1) the personally responsible citizen; (2) the participatory citizen; and (3) the justice-oriented citizen. This typology cannot account for the wide range and large number of activities but provides a good starting point to map the main types of citizens produced by environmental citizenship education.

The first type, the personally responsible citizen, is associated with education that teaches about the environment and attempts to build character and encourage students to assume individual responsibility regarding environmental issues. In general terms, this citizen aligns best with liberal conceptions of citizenship, mainly because of the emphasis on the individual. The second type, the participatory citizen, is a citizen that learns about the environment through activities and is encouraged to engage in collective, community-based environmental efforts through those activities. In general terms, this citizen aligns best with republican conceptions of citizenship, mainly because of the emphasis on participation and the collective approach to environmental action. The third type, the justice-oriented citizen, is associated with education that explains the social, economic and political dimensions of environmental issues, and teaches students to engage with those issues as part of their practice. In general terms, this citizen aligns best with ecosocial and insurgent conceptions of citizenship, mainly because of the emphasis on social justice.

Box 7.1

Top 10 eco education trends

The following trends on environmental education come from a list compiled by Kara Mitchell (2010: 6). She illustrates these trends with examples from Canada and the United States. However, similar activities can be found in other countries, mostly from the Global North. The trends are presented as innovations that are finding new ways to inspire students and the public, whilst clearing a path towards future sustainability.

1. Nature play. These activities seek to reconnect children with nature, addressing what the International Union for Conservation of Nature (IUCN) calls 'children's nature-deficit disorder'. www.iucn.org
2. Ecological footprint. Ecological footprint calculators allow students to measure the impact of their lifestyles on the planet and learn tips to reduce that impact. www.footprintnetwork.org
3. Climate Change Education. Public education campaigns are constantly developing and updating strategies for teaching this complex subject. www.theclimateproject.org
4. Food education.
5. Service learning.
6. Green schools.
7. Integrative science and integrated programmes.
8. Professional exchanges.
9. Learning vacations and ecotourism.
10. Spiritual environmentalism.

Student activity

Identify which of these activities you have experienced and/or seen in practice and examine what kind of green citizen was being constructed by them (e.g. personally responsible, participatory or justice-oriented).

These studies suggest that the dominant type of citizen being manufactured is the personally responsible citizen. The predominance of this portrait is hardly surprising if we consider the fact that since its early days 'the aims of environmental education were almost exclusively defined in terms of individual dispositions' (Postma 2006: 5). This is a citizen who is made aware of major environmental issues (e.g. biodiversity, climate change) and taught about the importance of preserving and promoting a healthy and ecologically balanced environment. Importantly, that information is not framed in terms of rights, that is, something we can demand from others or that nature can demand from us. Instead, it is framed primarily in terms of our individual responsibility to generate that type of environment – for the good of nature (biodiversity), our own well-being and that of future generations.

The participatory citizen ideal is fostered in different ways, most typically through in-class activities and school-wide projects that encourage student participation, such as recycling schemes, eco-art projects, conservation programmes, school and community gardens, field trips and collaborations (Figure 7.1). Scholars and theorists of environmental citizenship promote participatory education as the best pedagogy to generate the kind of active, responsible citizen seen as necessary to achieve transformative environmental sustainability. This approach has generated some interesting and notable initiatives (e.g. Houser 2005; Gebbels *et al.* 2011; Wake and Eames 2013). However, for the most part, the enactment of environmental citizenship continues to be associated with individual rather than collective actions, increasingly as consumer-citizens (as we shall see later in this section).

Figure 7.1 *The Goody Patch: The Goodwood Primary School Community Garden*

Source: Private Collection. Photographer: Seán Mullarkey.

The type of citizen that seems to be mostly absent in environmental citizenship education is the justice-oriented citizen. There is little evidence (at least in existing studies) suggesting that any significant attention is being paid to notions of social and environmental justice. To the extent that such citizens are being contemplated they tend to be in the context of teaching concepts such as fair trade (e.g. Pykett *et al.* 2010). But, on the whole, there is little evidence that justice is a significant dimension of education for environmental citizenship, at least not in the schools of the Global North. This point is underscored by a notable and interesting exception which comes from the Global South. In a study of environmental education in Costa Rica, Steven Locke shows how it is possible to integrate democratic values, citizenship education and environmental education in a single national curriculum that pays attention to the social, cultural and economic context. His study identifies several aspects that illustrate the cultivation of a fully rounded, justice-oriented (or at least justice-conscious), multicultural and ecological citizen. These aspects include: understanding the relationship between the community and its environment; addressing the social injustice of environmental degradation; and acknowledging and using traditional knowledge and practices that support ecological sustainability (Locke 2009).

In addition to the three types of citizens considered by Westheimer and Kahne, we must add a fourth and increasingly dominant type: the consumer-citizen. This is a citizen that shares many of the traits of the personally responsible citizen, and the active disposition of the participative citizen, but expresses these traits largely through consumption. In essence, consumer-citizens are active citizens who discharge their environmental responsibilities as individual green consumers. This type of citizen can also integrate aspects of the justice-oriented citizen (e.g. fair trade). In fact, the three types of citizen explored above are increasingly being integrated in the form of the consumer-citizen, irrespective of whether the focus is on individual responsibility, active participation or social justice. This emerging trend is one that allocates individual responsibility largely as consumer-citizens, demands participation largely as consumer-citizens, and to the extent that it does, provides for social justice through our actions as consumer-citizens.

The ascendancy of the consumer-citizen is perfectly illustrated in Carol Carter and Sarah Lyman Kravits' *Keys to Success: Cultural Awareness and Global Citizenship* (2013). This high school textbook focuses on four main questions: developing cultural awareness, building cultural competence, understanding global citizenship and taking action as a global citizen. The text ends with a section dedicated to 'Environmental Stewardship' that is dominated by the theme of consumer citizenship – with no reference to structures, government, corporations, social justice or political action. The key messages are: reduce your ecological footprint, consume responsibly and take action with your wallet. The textbook emphasizes individual responsibilities, primarily to the natural environment and to future generations, and portrays the citizen as a consumer – one that can best contribute by embracing initiatives related to waste reduction, recycling and sustainable consumption, i.e. the three Rs: Reduce, Reuse, Recycle. Students, and citizens in general, are increasingly trained in how best to do this by monitoring their ecological footprint.

The ecological footprint: ecological or neoliberal citizenship?

The ecological footprint is the main tool in the manufacturing of green citizens. The basic idea is simple: every individual, activity and population has an impact on Earth, through their use of natural resources and ecological services and the generation of waste. That impact is converted to a single indicator: the amount of bioproductive land and water required to produce the resources consumed and absorb the carbon dioxide emissions. This is the ecological footprint, which is expressed in global hectares (gha). The concept rapidly gained traction after the publication of Mathis Wackernagel and William Rees' book *Our Ecological Footprint* (1996). The ecological footprint was proposed by these authors as a simple, intuitive and practical method for measuring and comparing the sustainability of resource use amongst populations, as well as a policy guide and planning tool for sustainability. Over the years, different formulas have been used to calculate the ecological footprint of the entire planet, specific nations, cities and regions, as well as activities (e.g. tourism, agriculture), products, corporations, businesses, schools, universities and, last but not least, individuals.

WWF's Living Planet Report, published in 2010 but based on data from 2007, calculated the global ecological footprint as 18 billion hectares. This means that the Earth's people need 18 billion hectares of productive land in order to provide each and every person with the resources required to support their lifestyle and to absorb the wastes they produced. The bad news is that under prevailing technology and management systems there are only 12 billion global hectares available.

In other words, the report found that the Earth's people are using about 50 per cent more natural resources than the planet can regenerate. To put it differently, humanity is using the renewable resources of 1.5 Earths to meet our annual demands for energy, food, shelter and the things we do and buy. This means that we are living in a state of global 'overshoot' (or global ecological debt). The message is clear: we must reduce resource consumption if we are to live within the productive capacity of Earth. The report and plenty of other studies also note that there is a significant imbalance regarding the contribution to global ecological overshoot from the countries (and lifestyles) of the Global North and the Global South (Figure 7.2). These studies also suggest that 'society will need to meet the dual challenge of shrinking global demand and sharing this reduction in a way that is acceptable and viable for the entire global community' (Kitzes *et al.* 2008: 467).

There has been ongoing discussion regarding the suitability of the ecological footprint for planning and public policy, but there is widespread agreement regarding its political and educational power. Even those who criticize its scientific credentials acknowledge its power to 'raise public awareness and influence politics' (Van Kooten and Bulte 2000: 385) and agree that the ecological footprint is 'one of the most successful indicators for communicating the Earth's physical limits' (Aubauer 2011: 652). Indeed, from the start, the ecological footprint has been used widely in education, public policy and awareness campaigns, with an estimated 6 million people from over 45 countries accessing

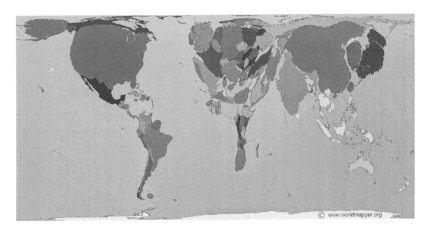

Figure 7.2 *World ecological footprint map*

Source: Worldmapper Project. Author: Benjamin D. Hennig.

online personal footprint calculators each year (Cordero *et al.* 2008). The question is: does learning about sustainability using the personal footprint calculators make a difference? And if so, does it manufacture a particular kind of citizen? The answer to both questions is 'yes'.

How does it work? The personal calculator asks a series of simple questions about shelter, food and transport, and then calculates the area of land required to support your lifestyle. The calculator typically produces a double score: the size of your ecological footprint on global hectares (gha); and the number of planets required if everybody lived like you. The current individual allocation of land per person is 1.8 gha. This means that if your footprint is larger than 1.8 gha your current lifestyle is unsustainable – and you owe an 'ecological debt' to those who are currently using less than their allocation (mostly people in the countries of the Global South).

The ecological footprint calculator is an excellent tool to get a sense of the general sustainability of our individual lifestyle and to identify which aspects contribute more to our footprint. Studies show that environmental education for sustainability based on the use of the ecological footprint 'enable students to evaluate, understand and examine the connection between personal and local lifestyle and the influence on ecological systems that support life on the global level' (Gottlieb *et al.* 2012: 197). In particular, personal calculators illustrate the significance of eating habits, which typically amount to around 50 per cent of the total individual footprint (Figure 7.3). The importance of food in personal ecological footprints has been reflected in the recent UN-led campaign 'Think, Eat, Save: Reduce Your Foodprint'. The ecological foodprint is impacted by a range of elements, but the main variable is the consumption of animal products, especially meat. However, containing the ecological impact of animal products is also a matter of production. The burden of industrial livestock or the 'grain-oilseed-livestock-complex' on the environment has been captured in the concept of 'ecological hoofprint' developed by Tony Weis (2013). In any case, ecological footprint calculators indicate that a change in eating habits (especially if we are frequent meat-eaters) can have a significant difference on the size of our ecological footprint.

However, not everyone agrees on the pedagogical effectiveness of individual ecological footprint calculators. Their impact seems to rely on the way in which they are used. In and of themselves, these calculators can have the opposite of the intended effect and discourage citizens from

Figure 7.3 *Individual ecological footprint (real example)*

improving their behaviour. This is largely to do with the fact that, because most people in developed countries live environmentally unsustainable lifestyles, the scores they produce and the feedback they provide are almost invariably negative. In fact, websites typically assume in their framing the scenario of overshoot and dedicate considerable space to providing tips for reducing personal footprints, such as: changing eating habits, driving and flying less, reducing water consumption (e.g. taking fewer or shorter showers, installing water tanks), reducing energy consumption (e.g. switching off lights and avoiding standby mode), installing energy-efficient appliances, shopping with care (e.g. taking your own bags, choosing products with minimal packaging, buying local products, growing your own fruit and vegetables), taking action (e.g. tracking business and government actions) and learning from others (e.g. checking case studies online).

The assumption seems to be that negative scores will inspire people to change their lifestyles in order to reduce their footprint. However, people can react to negative feedback in a variety of ways, including unproductive ones, such as:

> feeling overwhelmed, feeling that sustainability is impossible, giving up, justifying their unsustainable behavior, blaming other people for environmental problems, trivializing the impact of their own

behavior, deciding that environmental problems are not important, or discounting the accuracy of the footprint feedback itself.

(Brook 2011: 114)

Amara Brook even notes that one possible reaction is that people may respond by changing their environmental views to match their behaviour! Her study suggests that negative feedback can increase or decrease sustainable behaviour depending on pre-existing commitment to environmentalism. Thus, whilst negative feedback might inspire additional efforts from people with a strong environmental commitment, it can discourage those lacking such disposition from making significant efforts or even lead them to give up on their efforts altogether. This is particularly problematic when we consider that 'for the majority of available calculators, even when the most environmentally friendly options are adopted, one still exceeds the planet's biocapacity levels' (Franz and Papyrakis 2011: 395). These consistently negative scores can turn ecological footprint calculators into 'a doom-saying, off-putting instrument' that promotes the syndrome that the problem is 'so big nothing can be done' (Franz and Papyrakis 2011: 395). The only widely available calculator that seems to avoid this pitfall is the one provided by Redefining Progress.

The main reason personal footprints in the Global North are almost impossible to bring under 1.8 gha are the structural issues that escape our control as individuals, let alone as consumers. Individuals cannot change the transport system, the urban layout and the service infrastructure through consumption habits. Similarly, as individual consumers we have no control over large-scale industrial systems, such as food production, waste disposal and energy generation. In addition, affordability often means that we cannot decide where we live, how to travel between home and work and what types of houses we can buy, and we might even have little choice over what kind of food to purchase (e.g. if all we have access to is one or two large supermarkets). This tells us that, without significantly altering the social structure, physical infrastructure and industrial processes, individual changes and actions, while important, have a limited impact. In turn, this suggests that 'collective political action, not simply individual consumer action, is necessary to bring about the institutional changes that will be needed to achieve sustainability' (Obach 2009: 298).

The limits of consumer power – the need to go beyond consumption and shopping – are what inspired Annie Leonard to follow up her much celebrated short film about the impact of overconsumption on communities and the environment, *The Story of Stuff* (2007), with another one entitled *The Story of Change* (2012), and subtitled 'Why Citizens (Not Shoppers) Hold the Key to a Better World'. *The Story of Change* argues that we need to 'reengage our citizen muscles' and provides a three-step formula to relearn how to make change: a clear goal, teamwork and political action. The film advocates collective political activism rather than individual green shopping as the way to address environmental matters. In a recent interview, Leonard stated that, whilst there are plenty of things we can and should do on an individual scale, the real change comes from joining with others to demand action from leaders in government and business (Feeney 2013). In other words, we need to complement 'shrinking ecological footprints' with 'expanding political ones' (May 2009). These reflections suggest that the ecological footprint might be a problematic concept to promote justice-oriented notions of ecological citizenship, such as that theorized by Andrew Dobson (see Chapter 3).

Dobson proposes the individual ecological footprint as the political space of ecological citizenship – a space that would enable us (as citizens) to visualize and take responsibility for the impact that each and every one of us has on the planet. This should entail a sense of ecological debt from those with unsustainable ecological footprints (probably most of us) towards those with footprints smaller than 1.8 gha. In turn, this should generate a sense of ecological justice, both towards those who are currently using less than their share (typically because they live in poverty) and towards future generations (who will not be able to satisfy their basic needs or will inherit a depleted environment). However, the ecological footprint fails to adequately account for the past, that is, it cannot calculate ecological debt incurred through centuries of exploitation and appropriation of the environment (most notably by the Global North). This could imply that people in the Global North should reduce their ecological footprint below 1.8 gha, to pay our ecological debt and give additional ecological space to those in the Global South to develop (assuming that development is a desirable goal, and that it necessitates, at least temporarily, going over 1.8 gha per capita). But even if we were to ignore the past, present statistics are profoundly disturbing. For instance, if everyone on the planet were to enjoy the lifestyle of average US citizens, we would require four Earths! This means that to sustain their large ecological footprints the countries of the Global North run large ecological deficits with the Global South. The global rebalance demands radical change in the lifestyle habits of most people in the

Global North. The extent to which ecological footprint calculations can contribute to that change remains to be seen.

The issues raised by ecological footprint calculators should not discourage their use – especially not in educational settings. The key is to make sense of the findings in ways that account for contextual and structural issues and thus enable a critical interpretation of the results. On their own, ecological footprint calculators, with their focus on consumption and individual responsibility, tend to promote a neoliberal conception of environmental citizenship. This makes them an ideal instrument to illustrate the construction of the neoliberal environmental citizen (who acts on the environment as a consumer) and the limits of that conception of citizenship to address current environmental issues (as noted above). In fact, these calculators might be the best tool we have to engage critically with two of the most important conceptions of green citizenship: the most influential theoretical conception, ecological citizenship (as theorized by Dobson); and the increasingly dominant one, namely, neoliberal environmental citizenship. The extent to which ecological footprint calculators contribute to the manufacture of ecological and/or neoliberal citizens seems to depend largely on how users in general and educators in particular frame and interpret the results. But these interpretations do not take place in a vacuum. Instead, they take place in a context that is heavily influenced by popular culture and media representations. Indeed, the media provide both context and content when it comes to shaping our conceptions of environmental citizenship. Thus, it is appropriate that we now turn our final attention to those representations.

Green(ing) citizenship in media coverage and popular culture

Our lives are mediated. The media influence the way we perceive and make sense of the world. This, in turn, impacts on how we act in the world. In other words, the making of citizens cannot be fully understood without adequate consideration of the role of the media in shaping the context in which we are socialized as citizens. In this section we will explore the construction of environmental citizenship in the mass media, with particular attention to the impact of news media, documentaries and popular films in the making of green citizens. The majority of illustrations will come from the Global North, mainly because of the limited access to (and lack of studies of) representations of green citizenship in the Southern media. However, it is important to keep in mind that the Northern media in general, and Hollywood in particular, have achieved a level of global cultural penetration that makes their content relevant to the making of citizens, both in the Global North and the Global South (Miller *et al.* 2005). That level of penetration often reflects the transposition of particular green values and notions of citizenship from the North onto the South.

Framing the news: the ascendancy of infotainment

The media shapes the construction of environmental citizenship, first and foremost, through the general reporting of environmental issues and, particularly, through the coverage given to environmental activism. That impact derives in no small part from the framing of issues by the news media. The media in general and the news media in particular may not tell us what to think but they present us with a range of issues to think about, framed in particular ways. In other words, the news media 'provides us with the frames with which to assimilate and structure information about a whole range of social problems and issues' (Anderson 1997: 18). These frames are far from neutral or accidental. The production of news is a highly selective, controlled and constructed process, shaped amongst other factors by the medium, the sources, the reporters, the editors, news values, media laws and, last but not least, by media owners (Anderson 1997).

Media coverage of issues is contingent to a large extent upon who owns and controls the media, typically presenting the views of the establishment in general and, in the case of privately owned media, the particular views of media owners. In essence, media content tends to reflect the popular saying: 'he who pays the piper calls the tune'. In this sense, it is important to note the rise of the global media system following the neoliberal deregulatory policies of the 1980s. These policies have contributed to the convergence of media industries across traditional lines of production and the expansion of transnational media corporations. Some of these corporations have become global media empires, most notable the so-called 'holy trinity': Time Warner, Disney and News Corporation (McChesney 1999). These media empires control much of the production and distribution of information and entertainment that appears in newspapers, televisions and cinemas across the Global North and the Global South. The rise of the internet has introduced instability into the global media system, but corporations have begun to find ways to take over the infrastructure and exercise control of the content. To some extent, the internet is becoming part of the 'food chain' of corporate news media.

The other major aspect we need to consider when thinking about media coverage is the fact that the news media operates with particular conceptions of what is newsworthy. News media is event-centred, image-oriented and closely tied to a 24-hour daily news cycle. This means that environmental coverage is typically preoccupied with dramatic events with a strong visual component, and tied into a short time-frame that favours those groups who can provide timely press releases to feed the news cycle, most notably governments, think tanks, corporations and large NGOs. The event-centred nature of environmental reporting 'tends to divorce issues from their wider social and political context, and rarely incorporates a consideration of multiple explanations' (Anderson 1997: 134). The emphasis is on theatre and spectacle, images and catchwords, rather than meaning and substance. The blurring of the boundaries between reporting and spectacle is captured best in the concept of infotainment, the combination of information and entertainment that has become common in the news media, both in the Global North and the Global South (Thussu 2007). The influence of entertainment is also reflected in the term edutainment (the combination of educational content with entertainment value) first coined by The Walt Disney Company in 1948.

Nothing illustrates better the ascendancy of infotainment than the increasing presence of celebrities in all areas of the media, including the news. The lives of celebrities have become news in and of themselves and a regular conduit through which to cover other issues – including the environment. Indeed, celebrity has become a defining trait of our mediated societies. The impact of celebrity culture extends to politics in general and environmental politics in particular, both in terms of 'celebrification' and 'celebritization' (Driessens 2012). The first of the two terms refers to the transformation of ordinary people and public figures into celebrities (here: green celebrities). The second refers to the social and cultural embedding of celebrity in a range of social fields (here: environmental citizenship). The extent to which infotainment and celebritization bring attention to important environmental causes or distract audiences from our environmental predicament is a matter of debate, but there is no denying the increasing presence of celebrities in matters environmental. However, before we explore the celebritization of environmental citizenship, let us consider the impact of the main 'celebrity' associated with the media coverage of the environment – planet Earth.

The planet as cosmopolis: we are all Planeteers

The most direct and obvious impact of the media coverage of environmental issues regarding citizenship has been the promotion of global notions of citizenship, derived from the frequent use of images of planet Earth. These images declare that we are all Earthians – or, in the language of Captain Planet, we are all Planeteers. The emergence of Earth as 'the space of Ecopolitics' (Kuehls 1996) dates back to the release of the first photographs of the planet taken from space in the late 1960s. The first and most famous of these images, known as 'Earthrise', taken in 1968 by William Anders during the Apollo 8 mission, has been referred to as the most influential environmental photograph of all time, and is credited with having inspired the celebration of the first Earth Day in 1970. This and other early photographs, such as 'The Blue Marble', taken in 1972 by the crew of Apollo 17, helped to galvanize the environmental movement in the late 1960s and early 1970s (Heise 2008). Since then the media has used images of Earth from space as part of the regular coverage of environmental issues. These images have helped create the perception of the planet as the finite and fragile home of humanity, and have inspired the naming of subsequent global initiatives, such as Earth Hour (1997) and the Earth Charter (2000).

These images and initiatives invoke the notion of a global community and promote a sense of cosmopolitan citizenship. However, they also illustrate some of the limits and problems with the dominant global articulations of environmental citizenship. The images of planet Earth, which feature prominently in news media and documentaries, often serve to conceal inconvenient truths, such as, for example, the differentiated responsibilities of the Global North and the Global South for the current environmental predicament. In other words, the conception of global environmental citizenship promoted by these images obscures the asymmetry of different national (and individual) contributions to environmental degradation. Media invocations of global environmental citizenship ignore or gloss over history and relations of power, presenting everyone as equally vulnerable to environmental risks and equally responsible for preventing environmental degradation. These representations dominate the media coverage of climate change, but can also be found in pedagogical events like Earth Hour (Figure 7.4)

Figure 7.4 *Sydney Opera House lights out – Earth Hour 2013*

Source: WWF. Photographer: Simon Hewson.

and pedagogical texts such as *Captain Planet and the Planeteers* (see Box 7.2). The issue of global environmental justice is even absent from prominent activist and planetary documentaries such as Al Gore's *An Inconvenient Truth* (2006). The simple and often simplistic message is that 'we are all in this together' as equal, fellow Planeteers.

Green citizens: activists, celebrities, and consumers

The impact of the media on green citizenship derives, first and foremost, from their portrait of green citizens. Until recently, this was limited to depictions of green activists. However, in recent times, two additional figures have become increasingly significant: green celebrities and green consumers. These representations can be found in news media, documentaries and feature films from the Global North – all of which are increasingly globally available through the internet. The emergence of social media adds a layer of complexity to the media landscape but so far does not seem to have altered the main portraits – other than to enhance the visibility of green celebrities and accentuate the prominence of green consumers. These portraits contrast with images of insurgent citizens that are particularly prominent in media texts originating from

Box 7.2

Captain Planet and the Planeteers

The television series *Captain Planet and the Planeteers* is the best-known and most far-reaching pedagogical environmental animation. The series, created in 1990 by Ted Turner for his cable network TBS, screened a total of 117 episodes between 1990 and 1996. The cartoon still exists in syndication and continues to be shown regularly on US cable network Bloomberg, and around the world. The series won multiple awards, including two US environmental awards in 1993 and 1994. The influence of the series derives not just from its popularity but also from the Captain Planet Foundation, established in 1992, and the Planeteer Movement, launched in 2010. The movement was established so that fans can continue to connect to the show and apply its messages to their lives as real-life Planeteers.

The basic premise of the series is that Gaia (the spirit of planet Earth) calls upon five teenagers from around the world to lead the battle against the eco-villains wreaking havoc on the planet. Each of the Planeteers is given a magic ring which enables them to control one element of nature (Earth, Fire, Water, Wind) and one extra element, Heart. The five can join their forces to summon Captain Planet, the environmental superhero. Each episode ends with Captain Planet declaring to the audience: 'The Power is Yours!' The series is overtly pedagogical and environmental. Each episode ends with hints and tips for children (and viewers in general) to look after the environment.

Captain Planet and the Planeteers is rich with notions of citizenship and riddled with tensions. The series embraces cosmopolitan citizenship, as illustrated by the name of the superhero, Captain Planet. In fact, the character was designed to be as universal as possible, with each of the five Planeteers representing a region of the planet: Asia, Africa, Eastern Europe, North America and South America. Yet, qualities attributed to the different Planeteers, according to where they come from, are crude racial stereotypes. Moreover, despite his blue skin and green hair, Captain Planet is a classic white male all-American hero (King 1994). To this we must add the gendered and maternalistic representation of Gaia as Mother Earth. Thus, the seemingly multicultural formation betrays essentialist notions of identity and puts forward a classic image of the hegemonic universal citizen: a white, male, heterosexual, blue-eyed American. In other words: 'the Planeteers are a United Nations clearly led by the United States' (Sturgeon 2009: 115).

Resources

Captain Planet Foundation. Available at: http://captainplanetfoundation.org/
Captain Planet Intro HQ (published on 29 September 2013). Duration 1 min. Available at: www.youtube.com/watch?v=GyOMYC6mlsY
Captain Planet Trailer (published on 22 January 2013). Duration 3 mins. Available at: www.youtube.com/watch?v=_cIgKfReegg (fan-made)

the Global South, explored in the final segment of this section. We begin the exploration with the earliest of representations, that of green activists, as depicted in the media of the Global North.

Tree huggers and eco-terrorists: deluded and dangerous citizens

The portrait of green activists was (and often still is) largely negative, shifting between simplistic stereotypes of tree huggers and eco-terrorists. These depictions date back to the early days of the modern environmental movement. Images of the Chipko women facilitated the representation of environmentalism as a movement of 'tree huggers'. The expression was used to associate green activism with notions of soft, naïve and deluded politics. This image changed with the emergence of more direct-action groups, like Greenpeace, Earth First!, Sea Shepherd and the Earth Liberation Front (ELF). Suddenly, green activists began to attract the label of 'eco-terrorists'. Their activism came to be associated with violent politics and dangerous actions, albeit still driven by deluded ideals. The two images sometimes combined to portray green activists as 'dangerous clowns' – as was the case with the treatment of the ELF in *The New York Times* (Joosse 2012). This coverage displays a curious tension between the trivialization of environmental activism (the tree-hugger image) and its treatment as a serious threat (the eco-terrorist image).

Depictions of green activists as eco-terrorists have become normal, even in the absence of violent actions (which are very exceptional and almost invariably directed at property). The media seem unable or unwilling to accept any form of ecological politics that clashes with (or cannot be accommodated as part of) the free-market ideals and the economic interests of corporate capitalism – arguably a reflection of the fact that most media are in corporate hands. There is no space for civil disobedience let alone for property damage. The green citizen who wants to be a 'good citizen' cannot be a green activist. The mainstream media leaves little (if any) space for green politics outside of the liberal political framework of electoral politics. The citizen who wants to participate in the shaping of green policies must do so with their vote, or risk being portrayed as deluded and dangerous (violent, radical, terrorist). The most famous green activist who has attracted this label in recent times is Paul Watson, the founder of Sea Shepherd. Watson has 'embraced' and responded to this portrayal in a documentary film, directed by Peter Brown, entitled *Confessions of an Eco-Terrorist* (2010).

Similar representations of green activists can be found in popular animation series, such as *The Simpsons*, *South Park* and *American Dad!*. The figure of the tree hugger is present in all three series, although its treatment is somewhat sympathetic in *The Simpsons*, where Lisa Simpson emerges as an intelligent and lovable tree hugger. In many ways, Lisa is presented as a model green citizen. Indeed, her character has received several animal rights and environmental awards, including a special 'Board of Directors Ongoing Commitment Award' at the Environmental Media Awards, in 2001. By contrast, *South Park* and *American Dad!* take a dimmer view of green activists and combine the ridiculing of tree huggers with depictions of green activists as eco-terrorists. The figure of the tree hugger is particularly undercut in *American Dad!* through the character of Hayley. Her environmentalism is similar to Lisa's, but is treated as fake, mere posing and pretension. For its part, *South Park* combines the ridiculing of tree huggers with representations of green activists as eco-terrorists, in episodes such as 'Douche and Turd', 'Fun with Veal' and 'Free Willzyx'. These series also contain critical depictions of green consumers and green celebrities, but have no time for matters of ecosocial justice or the kind of insurgent citizenship found in media texts of the Global South.

Green celebrities and ecotourists: vicarious and adventurous citizens

The media coverage of green issues has become increasingly reliant on the presence and the manufacturing of green celebrities. The framing of nature and environmental issues through the images and actions of celebrities dates back to the 1980s, beginning with the emergence of 'green crusaders' and the ground-breaking work of *National Geographic*. The early crusaders were scientists made famous by their work with 'celebrity species' such as gorillas, pandas, rhinos and jaguars. They included figures such as Dianne Fossey, Jane Goodall and Cynthia Moss. Their actions were reported as those of 'morally enlightened individuals whose goal is to save a species from the ignorance, greed, and over population, of local people' (Vivanco 2002: 1199). The increasing influence of the entertainment industry has gradually transformed the early green crusaders into 'conservationist celebrities' whose profile sometimes resembles more that of an entertainment celebrity, as is the case with David Attenborough and the late Steve Irwin (Huggan 2013). This is also the case with Paul Watson, especially since his decision to film the reality show series *Whale Wars* (2008–2013) for the cable channel Animal Planet. In recent times, 'celebrity conservationists' have been joined, and somewhat superseded, by 'conservationist celebrities' (those who endorse various conservation aims) drawn from the entertainment industry, such as Leonardo Di Caprio, James Cameron, Sigourney Weaver, Robert Redford, Harrison Ford and Angelina Jolie.

The emergence of green celebrities as mediators between cultural trends and environmental issues has generated significant debate. Some have enthusiastically welcomed this development as a 'win, win, win!' (Koerner 2013). This optimism is also illustrated in the title of Tommi Lewis Tilden's article on *The Daily Green*: 'How Green Celebrities Helped Save Our Planet This Summer' (2008). Others are less euphoric and point to the many caveats that come with the celebritization of the environment (Goodman and Littler 2013). Either way, the increasing salience of green celebrities in the media coverage of environmental issues has significant implications for environmental citizenship. Specifically, celebrities tend to inspire ordinary citizens in ways that ultimately promote the consumption of nature and thus contribute to the manufacturing of neoliberal environmental citizenship, in two different (but ultimately complementary) forms: passive (vicarious citizenship) and active (adventure citizenship).

The experience of watching green celebrities in action offers spectators the feeling of being part of an effort to preserve the natural wonders displayed on their screens. These shows promote a form of vicarious activism that might translate into making a financial contribution to conservationist organizations supported by particular celebrities. In addition, they inspire affluent citizens to tour the sites covered in the shows, helping to fuel the nature tourist industry that used to be associated with safaris, and has been redesigned, reinvented or simply rebranded as ecotourism (Vivanco 2004). In other words, these celebrities inspire passive supporters, who defend the rights of nature from the couch (i.e. vicarious citizenship), and active consumers, who become the clientele of the hotels and resorts created by the tourist industry to enable the 'authentic' experience of the ecotourist (i.e. adventurous citizenship). These two articulations of environmental citizenship share critical characteristics: they are cosmopolitan (focused on the planet as the territorial referent) and neoliberal (focused on the consumption of nature, through televisual spectacles and tourist adventures). Moreover, they promote the conservation agenda driven by the Global North, deepening the 'right to nature' of the Global North over the 'natural spaces' of the Global South.

Green consumers: sensible and sustainable citizens

The increasingly prevalent representation of green citizens in the media is that of the sustainable, green consumer. This image has come to replace the focus on green activists. Green consumer-citizens do not get involved in politics, not even electoral politics. Instead, they embrace the ideal of sustainability and take up the challenge through sustainable consumption, changing consumer practices and habits in small and reasonable ways, voluntarily and without making a point of displaying or advocating particular lifestyles. This image of the green consumer can be found across news stories, documentaries and newspaper supplements dedicated to promoting green products and technologies (e.g. e-bikes, hybrid cars, solar panels, etc.) and providing tips on how to save energy and reduce waste. The narrative of sustainable consumption works much in the same way as that of corporate social responsibility. They are both presented as win-win scenarios, that is, practices that save money while helping to 'save the planet'. This narrative reflects a neoliberal conception of environmental citizenship – one that has come to prevail across all media platforms, including public service television in countries like Germany and Great Britain (Inthorn and Reder 2011).

The neoliberal approach to environmental citizenship is best illustrated by films and series dedicated to encouraging audiences to make changes in their lifestyles and patterns of consumption. The most famous example is arguably Laura Gabbert and Justin Schein's documentary *No Impact Man* (2009). This film follows Colin Beavan and his family during their year-long project to live with zero environmental impact (Figure 7.5). The experiment results in the family adopting a frugal way of life that includes deprivations such as foregoing takeout food and toilet paper, avoiding out-of-season produce, reducing the consumption of coffee, foregoing most modes of transportation including elevators, and eventually going without electricity altogether. The message is clear and simple: saving the planet one family at a time.

The concept of 'sustainable households' has also been taken to television in Australia, where the public multicultural broadcaster SBS produced the series *Eco House Challenge* (2007). The premise of the series is similar to that behind *No Impact Man* – teach households ways to minimize their ecological footprint through sustainable consumption and the reduction of wasteful consumption. The show results in a series of 'experiments in environmental citizenship' that illustrate how 'the management of populations in late liberal societies is increasingly occurring at the level

Figure 7.5 *The Beavans* – No Impact Man *(2009)*

Source: Photograph courtesy of Colin Beavan.

of everyday life and consumption through a focus on the conduct and lifestyles of individuals' (Lewis 2012: 324). These media texts are part of a trend to adopt the reality television format as a pedagogical tool to promote Do-It-Yourself (DIY) citizenship (Ouellette and Hay 2008). The extent to which these and similar programmes deepen neoliberal citizenship or can emancipate citizens is still an open question, but in the overall climate they seem to fit neatly with the neoliberal agenda, echoing the signature call of Captain Planet: The Power is Yours!

The picture of environmental citizenship we convey to today's children is dominated by concepts of consumer citizenship (e.g. sustainable consumption). This is particularly visible in children's animation – a genre shaped largely by the cultural hegemony of Hollywood. This genre has produced an explosion of environmental texts in recent times, including: *FernGully* (1992), *Free Willy* (1993), *Finding Nemo* (2003), *Happy Feet* (2006), *Wall-E* (2008) and *The Lorax* (2012). These and similar texts invite children (and parents) to care for and act on behalf of the environment. But, crucially, the citizen is interpellated as a consumer (an economic animal, or *homo economicus*), not as the political animal (or *zoon politikon*) traditionally associated with the portrait of the citizen. In addition, these films ignore collective action and government regulation as part of the solution to our environmental predicament. The relevant agents are citizens (as consumers) and corporations. The individualized response to environmental degradation in these films denies any need to change institutional structures that drive or enable the consumerism they sometimes overtly criticize. In fact, most of these films promote a culture of consumption through the relentless production and advertising of toys and other merchandise associated with the films. In any case, this emphasis on individual consumption and individual responsibility contrasts markedly with the typical representations of environmental citizenship in the media texts from the Global South.

Insurgent citizenship:
the politics of survival, dignity and respect

The dominant approach to environmental issues in the media texts of the Global South is significantly different to its approach in those from the Global North. The main differences are: their focus on collective rights (rather than individual duties), their treatment of nature as a commons (rather than a source of commodities) and their defence of the natural commons as a public good (rather than through their privatization). These aspects are interrelated and moulded into narratives that often bring together environment and development. Similarly, the media coverage of environmental activism in the Global South contrasts significantly with the one that dominates in the media coverage of the Global North. The focus on the individual, personal duties and lifestyle issues that dominates the coverage in the Global North contrasts with focus on the collective, human rights and survival issues typical in the media texts of the Global South. In general terms, the portrait of environmental citizenship that emerges from these narratives resembles the multidimensional articulations of environmental citizenship and the concept of insurgent citizenship we explored in Chapter 4.

These themes are particularly well illustrated in the numerous documentaries about the struggles against the Narmada dams in India, such as Sanjay Kak's *Words on Water* (2002), and the 'water wars' of the 1990s in Cochabamba (Bolivia), such as Tin Dirdamal's *Rivers of Men* (2011). The struggle against the privatization of water in Cochabamba led to one of the most famous citizen revolts against the extension of the neoliberal paradigm to the Global South. The conflict between government and corporations on the one hand, and citizens (especially the indigenous population) on the other, was portrayed as the continuation of the colonial process in Icíar Bollaín's *También la lluvia* (2010). This film tells the story of a film crew that arrives in Cochabamba in 1992 to produce a film depicting Columbus' first voyage to the 'New World' in 1492. The title (literally, Even the Rain) refers to the fact that even the privatization of rainwater was contemplated at the time. This film is a clear metaphor that portrays corporations as the new colonizers and condemns the neoliberal approach to nature, asserting instead the human right to water within a framework of nature as commons that contrasts radically with the dominant conception of nature in the media of the Global North.

The politics of survival comes through in particularly compelling and original fashion in Lucy Walker's documentary *Waste Land* (2010). This film follows the Brazilian artist Vik Muniz, an established artist working in New York, upon his return to Brazil to photograph *catadores* or waste-pickers – the people who make their living from picking through and recycling materials from the waste garbage dumps that lie on the outskirts of cities. Muniz, in collaboration with the *catadores*, begins to create a series of monumental photographic portraits of the *catadores*, made up of collages of carefully arranged items of garbage. The artistic result is truly astonishing, but what is more astonishing is how the film transforms the 'indignity' of picking through the rubbish into a tale of human dignity contained within and emerging from the world's largest garbage dump, Jardim Gramacho in Rio de Janeiro.

The people depicted in the documentary represent a different kind of insurgent environmental citizenship. Their 'insurgency' resides in the dignity and strength with which they 'embrace' their current predicament, one that places them in the middle of tons of garbage generated, not exclusively, but disproportionately, by the urban affluent classes. The documentary puts names and faces to those who find themselves at the extreme end of supply-chain citizenship. Their labour is both the product of the 'brown economy' that generates the waste they end up sorting through, and also, paradoxically, a compelling illustration of a different sort of 'green economy' – the one they create through their work as garbage sorters and recyclers. This is a different (and rarely seen) articulation of environmental citizenship. These are the green citizens who silently demand social and environmental justice, while they continue to work tirelessly on the wastelands of the Global South, filled with the rubbish that is generated by the reckless overconsumption of the Global North, and by those who increasingly enjoy similarly affluent lives in the Global South.

The politics of dignity are also at the heart of Denis Chapon's *Abuela Grillo* (2009). This short film, based on an indigenous story, resulted from the collaboration between Bolivian animators and the Animation Workshop of Denmark. The film tells the story of Abuela Grillo, or Grandmother Cricket. The story is as simple as it is powerful. When Abuela Grillo sings, it rains. But one day, after being mistreated and humiliated by the villagers, she leaves for the city. Once there, ruthless capitalists discover her extraordinary ability. They lock her in a basement and torture her to make her cry, and then they bottle and sell her tears. The scheme brings disastrous results to the city. Eventually, the people and Abuela Grillo revolt against her rapacious torturers and the state authorities that support them. *Abuela Grillo* is a clear invocation of the water wars in Cochabamba, and reflects the multidimensional conception of environmental citizenship that is often typical of the Global South. The film brings into the picture citizens, governments and corporations, treats nature as commons, and involves collective political activism. In addition, the figure of Abuela Grillo brings a strong gender dimension (e.g. old women's knowledge), and the indigenous origins of the myth invoke the notion of indigenous citizenship. In short, *Abuela Grillo* illustrates that other representations are possible, but also (by its very exceptionality) highlights the strength of the current dominance of neoliberal representations of environmental citizenship.

Despite its indigenous origins and inspiration, *Abuela Grillo* speaks of universal themes, particularly that of human dignity. Indeed, while the tale is a clear representation of the water wars and thus evokes, above all, the 'right to water', the heart of the tale is not to be found in matters of rights, duties or even identity, but in human dignity, a concept that goes beyond the legalistic language of human rights. Dignity is what makes Abuela Grillo turn her back on the humiliations suffered in the village, and then rebel against the exploitation she is subjected to in the city. In a sense, the content of this film cannot be contained in the language of citizenship, environmental or otherwise (although it clearly brings up plenty of elements that relate to it), but must be interpreted as a tale of human dignity, but also of the dignity of nature. Abuela Grillo is, after all, a cricket! By contrast, the rest of the characters in the film are humans. Thus, the film is an indictment of the humiliation and exploitation of humans and nature (both in the rural and the urban environments). In short, *Abuela Grillo* places dignity at the centre of its portrait of insurgent citizenship, but this time extended to the entire natural world.

The politics of dignity at the heart of texts like *Waste Land* and *Abuela Grillo* call for notions of environmental citizenship that go beyond the numerical calculations of ecological footprints, and bring into the picture the dignity of people and nature, and the respect that such dignity demands. In doing so, these texts suggest that we complement the three Rs of sustainable consumption and environmental citizenship (Reduce, Recycle, Reuse) with a fourth and perhaps more important R: Respect.

Summary points

- Citizens are made, not born. The making of citizens is a long and complex process of political socialization that takes place through formal education, news media and popular culture.
- The dominant model of citizen being manufactured by formal education is a globalized personally responsible citizen, with responsibilities towards the environment and future generations, but not well attuned to global social and environmental injustices.
- The most popular tool to promote environmental citizenship is the ecological footprint. Ecological footprint calculators promote the privatization of responsibilities and the identity of the consumer-citizen but, when used critically, they can also reveal the limits of neoliberal citizenship.
- The media present particular images regarding the appropriate role for citizens, governments and corporations in relation to the environment. The green consumer has replaced the green activist as the dominant representation of the green citizen in the media of the Global North.

• Neoliberal conceptions of environmental citizenship are directly challenged by ecosocial and justice-oriented representations of insurgent citizens in media texts originating in the Global South. These often emphasize the participative, collective aspects of citizenship, the connection with other forms of identity and knowledge, and transformative attitudes that challenge capitalist power relations, and demand respect for the dignity of humans and nature.

Selected readings

Bell, D. (2004) 'Creating Green Citizens? Political Liberalism and Environmental Education'. *Journal of Philosophy of Education*, 38(1): 37–53.

Brook, A. (2011) 'Ecological Footprint Feedback: Motivating or Discouraging?' *Social Influence*, 6(2): 113–128.

Gough, S. and W. Scott (2006) 'Promoting Environmental Citizenship through Learning: Toward a Theory of Change', in A. Dobson and D. Bell (eds.), *Environmental Citizenship*. Cambridge, MA: The MIT Press, 263–285.

Hayward, B. (2012) *Children, Citizenship and Environment: Nurturing a Democratic Imagination in a Changing World*. London: Routledge.

Inthorn, S. and M. Reder (2011) 'Discourses of Environmental Citizenship: How Television Teaches Us to Be Green'. *International Journal of Media and Cultural Politics*, 7(1): 37–54.

Maniates, M. F. (2001) 'Individualization: Plant a Tree, Buy a Bike, Save the World?' *Global Environmental Politics*, 1(3): 31–52.

Nowak, A., H. Hale, J. Lindholm and E. Strausser (2009) 'The Story of Stuff: Increasing Environmental Citizenship'. *American Journal of Health Education*, 40(6): 346–354.

Wackernagel, M. and W. Rees (1996) *Our Ecological Footprint: Reducing Human Impact on the Earth*. Gabriola Island, BC: New Society Publishers.

Online resources

Abuela Grillo (2009) Dir: Denis Chapon. Duration 12.42 mins. Available at: http://vimeo.com/11429985. Also available at: www.youtube.com/watch?v=AXz4XPuB_BM

Green Ninja: Footprint Renovation (uploaded 10 June 2011). Duration 2.48 mins. A climate action superhero comes to the rescue of a man with a gigantic carbon footprint. Available at: www.youtube.com/watch?v=UeYOZgbgG1Q

I Am the Earth (2008) (uploaded 28 June 2011). Duration 3.56 mins. Available at: www.youtube.com/watch?v=Wkn1B7cJYQo

Inspiring Environmental Citizenship (published on June 21, 2012). Duration 5.05 mins. Film about learning escapes inspiring environmental citizenship through outdoor learning. Available at: www.youtube.com/watch?v=dnBEHxVGUhQ

No Impact Man: Official Trailer (2009) (uploaded 29 July 2009). Duration 2.29 mins. Available at: www.youtube.com/watch?v=Z9Ctt7FGFBo

No Impact Man: The Documentary (2009) Dir: Laura Gabbert and Justin Schein. Duration 93 mins. Available at: www.filmsforaction.org/watch/no_impact_man/

The Story of Change (2012) Official Version. Duration 6.29 mins. Can shopping save the world? *The Story of Change* urges viewers to put down their credit cards and start exercising their citizen muscles to build a more sustainable, just and fulfilling world. Available at: www.storyofstuff.org/movies-all/story-of-change/

Waste Land (2010) Dir: Lucy Walker. Duration 100 mins. Trailer 2.16 mins. Available at: www.wastelandmovie.com/

WWF: Living Planet Report (2014). Available at: http://wwf.panda.org/about_our_earth/all_publications/living_planet_report/. See also: Living Planet Report (2012). Available at: http://wwf.panda.org/about_our_earth/all_publications/living_planet_report/2012_lpr/

Student activity

Calculate and discuss your ecological footprint. The most widely available personal calculator can be found on the websites of the WWF and the Global Footprint Network. Here is the link (if it does not work, simple search for 'ecological footprint calculator'): http://footprintnetwork.org/en/index.php/GFN/page/calculators/

The process of measuring our own ecological footprint helps us to better understand how our everyday choices and activities contribute to our ecological impact. It also helps us to identify what activities are having the biggest impact on the environment and inspire us to take personal and collective actions to reduce our impact and live within the means of one planet. The calculator asks a series of simple questions and then calculates the area of land required to support our lifestyles.

Keep in mind the issues raised about the ecological footprint in the relevant sections of the textbook. Once you have calculated and reflected on your ecological footprint, join Moobius in 'The Meatrix'. This is a hilarious animation that highlights the problems of factory farming: www.themeatrix.com. You can also check the video 'Green Ninja: Footprint Renovation', which brings the concept of the ecological footprint into the realm of popular culture, by creating a green superhero that some have called 'the new Captain Planet'.

Discussion questions

- Which of the four types of citizen explored in this chapter (i.e. responsible, participatory, justice-oriented and consumer) was predominant in the environmental education you received at school?
- Do you think we can live in an environmentally sustainable way without attending to social justice? Why or why not?
- How much of your sense of citizenship (environmental or otherwise) comes from media coverage of representations of citizenship?
- Do you think dignity is something that needs to be included in our conceptions of citizenship, or do you find that this is too abstract and/ or intangible a concept to be included?

Conclusion
Into the future

Anyone who isn't confused really doesn't understand the situation.
Edward Murrow (1969)

There can be no doubt left as to the diversity and complexity of intersections between citizenship and the environment over the course of the past half a century, but particularly since the 1990s. The extent to which this text manages to map and make sense of those intersections I leave for you, the reader, to judge. In any case, the above quote is not meant as an excuse for (hopefully not!) failing miserably in that attempt. The truth is that anyone who is not at least partially confused does not fully understand the complexity of the ongoing relationship between citizenship and the environment – a relationship that is fluid and dynamic and will continue to evolve and develop into the future. Indeed, as we will see towards the end of this conclusion, there are several emerging issues that promise to complicate even further the picture of environmental citizenship in the years ahead. But before we take a look at those issues, we will revisit the main aspects explored in the textbook, highlighting some of the key arguments and dynamics that shape environmental citizenship across the world.

Differences and convergences

This book reveals how the historical resilience of citizenship is currently being tested, perhaps like never before, by the intrusion of the environment into the realm of politics in general and citizenship in particular. Environmental concerns and ecological values are challenging modern traits of citizenship (e.g. its exclusive association with the nation-state) and even some its essential historical attributes (e.g. its exclusive application to humans). The early chapters outline and discuss these challenges in the theoretical realm. For the most part, we saw how theorists from different political traditions are finding ways of rethinking citizenship that address these challenges, retaining the basic elements (membership, rights and duties) and adjusting their content in ways that are fundamentally compatible with traditional conceptions of citizenship (e.g. liberal and republican). Others, however, view those challenges in a different light and are developing novel conceptions of environmental citizenship, sometimes radically different from conventional ones. The most influential of these novel articulations is the work of Andrew Dobson, particularly his concept of ecological citizenship, developed in his seminal text *Citizenship and the Environment* (2003).

The cover of Dobson's book suggests that the impact of the environment on citizenship is akin to that of the French Revolution. Whilst the ultimate impact of the environment on citizenship is yet to be determined, there is little doubt that something revolutionary can happen to citizenship if some of the most radical proposals, particularly those driven by ecocentric values, come to fruition. The different nature – no pun intended – of some of these challenges forces us to question whether the outcome of such changes could even be called citizenship. The inclusion of future generations, let alone animals and nature in general, would signify in the eyes of many the end of citizenship as we know it, or would demand the use of a different term altogether. So, if the talk of revolution is warranted, and if this revolution was triumphant, would it mean the end of citizenship? Or only the end of citizenship as we know it? And either way, to paraphrase the REM song, should I feel fine?

These questions cannot be answered at present, but might be decided in the course of this century. The answers will probably be determined in the realm of practice, that is, by the interplay between the actions of citizens, government policies and business initiatives around the world. These actions, policies and initiatives are the content we explored in the central chapters of the book. These chapters reveal the diversity and complexity of practical manifestations of environmental citizenship, reflecting and/ or producing different theoretical articulations in the process. In addition, these chapters illustrate the difficulties of turning ideas into practice and, in turn, reveal the need for theories to keep up with the fluid and dynamic realm of practice. Moreover, this part reveals significant differences as well as important points of convergence between the Global North and the Global South.

The differences regarding the practical manifestations of environmental citizenship around the world derive largely from contrasting political and material contexts. As stated several times: context is all. Thus, for example, we noticed how affluence and poverty shape environmental priorities, but not always or necessarily in ways we might expect. Indeed, contrary to early (and still widespread) views that consider the environment a concern of the affluent, we have seen how some of the most revolutionary articulations of environmental citizenship originate in countries of the Global South. We have also seen how environmental movements in the Global South operate with and produce more comprehensive and multidimensional conceptions of environmental citizenship that have social justice, democracy and equality at their heart – aspects often seemingly taken for granted in the Global North but essential to the experience of citizenship in the Global South. This wider conception of environmental citizenship is something we might all do well to embrace, or risk having some of our present liberties and equalities compromised, particularly if we uncritically embrace the discourse of individual environmental responsibility currently being promoted around the world, especially in and by the Global North. To be sure, there is nothing intrinsically wrong with assuming personal responsibility for the environment, but we must guard against the use of this discourse to distract from government and corporate responsibilities and/or shut down debate about structural dynamics and global injustices.

The contrast between the manifestations of environmental citizenship in the Global North and the Global South derives also from the different ways in which human–nature relations are typically articulated. The practical articulations of environmental citizenship in the Global South tend to operate with a view of the relation between humans and nature that attaches intrinsic value to nature but does not neglect the value of human life. They regard all life as something to be respected, that is, as something worth preserving with dignity. In other words, the discourses that underpin those articulations contrast significantly with the dominant discourses in the Global North. The discourses of the Global South are often anthropocentric when compared with the ecocentrism of deep ecologists and conservationists of the Global North, whilst appearing ecocentric when compared with the anthropocentrism of ecological modernists and liberal environmentalists of the Global North.

The different practical and discursive manifestations of environmental citizenship in the Global North and the Global South also come through in the practices and discourses of transnational corporations. Whilst their embrace of environmental citizenship is framed in global terms (e.g. the Global Compact), their actions betray the double standards with which they often operate in the Global North and the Global South. These are partly a reflection of the different material and political contexts in which they function – contexts of which they take advantage, often with the acquiescence and sometimes at the invitation of the political authorities of the Global South. In addition, giant retail corporations are setting up schemes that outsource their green duties to their suppliers. Their power effectively translates into the imposition of environmental responsibilities from the Global North onto the Global South.

Yet, despite all these differences, these chapters have also identified patterns of global convergence in the practical manifestations of environmental citizenship. Some of the convergences derive from the increasing collaboration between activists from all around the world, forming networks and promoting notions of global and glocal environmental citizenship. These networks and initiatives do not necessitate the elimination of cultural differences, but do call for basic standards of environmental human rights and well-being for *all* humans, including future generations. There is also significant convergence in the way governments are incorporating the environment into legal (often constitutional) frameworks, as well as regarding the integration of environmental citizenship in frameworks of governance and governmentality. The proliferation of environmental governance is well documented, but the more interesting and arguably more disturbing trend is the globalization of governmentality, which seems to be gradually extending across the Global North and, driven by international institutions and regimes, into the Global South.

The convergence around governmentality introduces problematic dynamics into the future practice of environmental citizenship. Of course, if we take citizenship to be a mechanism to manage populations, then environmental citizenship is only doing what citizenship has been doing all along, that is, serving to establish social, political and economic order through the mechanisms of inclusion/exclusion, only now in the realm of the environment. In fact, the total inclusion of humans (present and future) and natures (human and nonhuman) could be the most efficient and palatable way to govern populations towards a sustainable future in order to bring about the required changes – particularly in the face of suggestions by prominent ecologists to 'put democracy on hold for a while' (Lovelock, cited in Hickman 2010). In other words, environmentality could emerge as the best governmental approach to produce sustainability in a context where we need urgent change and we value freedom and democracy. Yet the price attached to this environmentality might be transparent democracy, and ultimately freedom itself. Besides, governmentality is not immune to context. Different contexts will shape governmentality in different ways. So the 'solution' is not as simple as adopting this approach to the governing of the environment.

The ascendancy of governmentality is taking place in a neoliberal context that produces environmental citizenship as a form of neoliberal environmentality. The push in this direction comes particularly strongly from corporations. The embrace of environmental citizenship by the corporate sector is certainly astonishing, and upon close (or not so close) inspection clearly self-interested. The significance of the corporate embrace cannot be overstated. This is indeed a century in which our environmental values have become 'increasingly mediated through the corporate sector' (Whitehead 2011: 4). The neoliberal turn has also translated into a rapid shift towards the language of personal and corporate responsibility through self-regulation, and a shift towards economics in general (e.g. market rules and values) and consumption in particular (e.g. sustainable consumption) in articulations of environmental citizenship. 'The agenda' – as Gill Seyfang notes – 'has narrowed from initial possibilities of redefining prosperity and wealth and radically transforming lifestyles, to a focus on improving resource productivity and marketing "green" or "ethical" products' (2005: 294). This narrowing of the agenda is also noticeable in the pedagogical and media frames manufacturing green citizens, present and future.

Neoliberal citizenship, environmental or otherwise, might not be the dominant theory (yet) and it might not be the dominant practice either (yet), but it is clearly the model that presently dominates the manufacturing of citizens, especially through the mass media, and increasingly through schooling. Neoliberal citizenship is what we are predominately teaching the future (green and global) citizens, especially in the Global North. Increasingly, the picture of environmental citizenship we convey to today's children is dominated by concepts of consumer citizenship (e.g. sustainable consumption). This is noticeable when we examine the dominant representations of green citizens in mass media and popular culture, and is particularly visible in children's environmental animation. These texts invite children and parents to care for and act on behalf of the environment, but viewers are interpellated as consumers, or economic beings (*homo economicus*), not the political animals (*zoon politikon*) traditionally associated with the portrait of the citizen. In any case, the future of environmental citizenship will be determined not just by the citizens we manufacture and the citizens we become, but also by the themes, challenges and opportunities that will emerge in the years ahead. Some of the most significant are: population growth, animal rights and new technologies.

Emerging issues and themes

The concern with a growing population is making a powerful comeback, with references to the 'population bomb' and the 'population tsunami' becoming once again common, although the issue never went completely away. Of course, population is a legitimate concern when it comes to determining the carrying capacity of Earth. But over the years it has become clear that the main issue when it comes to sustainability is human lifestyles, and particularly the consumer (meat-eating) lifestyles typical of the Global North. However, as the global population grows and more people and nations embrace that lifestyle, the 'numbers game' becomes real and critical. In this context, population control might reemerge as a legitimate proposition to address our environmental predicament and promote environmental sustainability. The danger is that, in the process, environmental justice and reproductive rights will be undermined, particularly in the Global South (Di Chiro 2008). Indeed, this issue can have a profound impact on citizenship, mainly through the clash between the consumer rights of the affluent (who average fewer children) and the family and reproductive rights of the poor (who average more children), but also between human rights and animal rights and the rights of nature in general. Ironically, the ecological footprint – rather than promoting justice – can bring up the issue of population as central to the future of environmental citizenship. Indeed, the ecological footprint facilitates viewing ecological sustainability as a numbers game, since a reduced population would allow the expansion of the 1.8 gha per capita currently at our disposal. In any case, the numbers game will not go away easily, especially considering the population will continue to grow until it reaches some 9 billion.

The issue that will probably bring population into the discussion of environmental citizenship and citizenship in general in a very direct and explicit way is that of so-called climate refugees. This term is used to refer to people displaced from their homes, and sometimes their countries, because of climate change, particularly ocean level rise. In cases where the displacement is internal, the issues will be less challenging to the concept of citizenship, but in cases where the issues force people to relocate to a new country, issues of membership, cultural identity, rights and duties will complicate citizenship in significant ways, and will probably fuel the concerns with population growth in poor low-land areas of the planet, such as Bangladesh. The people who manufacture, often under appalling conditions, the cheap t-shirts and clothes that the planet in general and the Global North in particular buys in unsustainable ways, will be 'asked' to have smaller families . . . but continue to produce cheap clothes.

The other issue that will create some uncomfortable and challenging issues for environmental citizenship is the increasing tendency of the animal rights movement to recognize animals in general and domestic animals in particular as legal persons, and in some cases as 'animal citizens'. The idea of bringing 'animals into citizenship' can be traced back to the animals rights advocacy of Alice Morgan Wright and Edith Goode in the 1940s (Birke 2000), and best exemplified at present by the Great Ape Project and the Nonhuman Rights Project. The concept of animals as 'legal citizens' has gained further traction with calls to extend citizenship to domestic and domesticated animals. These include, for example, David Grimm's *Citizen Canine* (2014) and, most significantly, given the academic pedigree of its authors, Sue Donaldson and Will Kymlicka's *Zoopolis* (2011), a political theory of animal rights that could have a profound influence on our conception of citizenship in the decades ahead, perhaps as significant as Kymlicka's theory of minority rights, articulated in his ground-breaking text *Multicultural Citizenship* (1995). Their work on 'animal citizenship' sketches an argument for why 'justice requires the extension of citizenship to domesticated animals, above and beyond compassionate care, stewardship or universal basic rights' (Kymlicka and Donaldson 2014: 201).

These initiatives can deepen the tension between human rights and animal rights in general, but also feed into the rights gap between the Global North and the Global South. Indeed, animal protection and population control have often been pursued at the same time. Animal citizenship, in a context where the ecological footprint of the domestic animals of the Global North is often larger than that of humans in the Global South, could easily become enmeshed in the discussion about the consumer lifestyles of the Global North and the population growth in the Global South. The upshot being that the 'animal citizens' of the Global North could end up with more legal protection, and indirect political agency through the political and consumer power of their owners (and fellow citizens), than the people of the Global South.

The third and final issue that could have significant impact on the future of environmental citizenship is technology. Many hope that technological advances will solve the numbers game by producing what we need more efficiently and with less cost to the environment. But in terms of citizenship the most interesting technology is digital technology, particularly due to its potential to enhance citizenship participation, including environmental activism. Indeed, some have begun to use the expression 'digital tree-sitting' and others are exploring the question of whether online environmental activism can deliver change offline. In addition, digital technology can enhance the sense of global citizenship and provide further opportunities to come up with shared solutions to our environmental predicament. Some even hope it will reduce the environmental impact of our lifestyles – through, for example, the manufacturing of houses (and other goods) with 3-D printers. But digital technology has its own ecological footprint. Moreover, the online world can distance us from nature, making us even more alien to our natural surroundings than we already are, particularly in the Global North. And this is not to mention the impact that some new technologies (particularly biotechnologies) can have on our understanding of what it means to be human, let alone to be a citizen. These issues alone indicate how profoundly 'confusing' the field of environmental citizenship still is and will continue to be, even as we try to make sense of it as it currently stands.

In any case, none of these issues alter (at least not yet) the fundamental picture of environmental citizenship presented in this text, a picture dominated by the neoliberal discourse that currently shapes the global articulation of citizenship in general, and of environmental citizenship in particular. How these issues will impact upon or interact with neoliberal and other notions of citizenship remains to be seen. The extent to which the neoliberal approach to citizenship will prevail in practice (and perhaps, eventually, in theory) is impossible to predict. Classical conceptions of citizenship have lasted for centuries and powerful alternative ones have emerged in recent years. The different models will continue to coexist in tension into the future. But along the way some will prevail over others, at different times and in different places. And this poses plenty of fascinating questions.

Will neoliberal environmental citizenship extend across the Global South, especially as its peoples begin to enjoy levels of affluence similar to those in the Global North? Will ecosocial conceptions of citizenship gain traction in the Global North? Will concerns with justice and democracy become secondary to sustainability? Or will they become central to the models of citizenship we create to address environmental concerns? Can green shopping be compatible with or even promote ecological citizenship? Can corporations and consumers generate a sustainable world in the absence of government regulation? Critics note that the discourse of sustainable consumption (at least implicitly) allows for individuals, governments and corporations alike to choose an 'easy symbolic alternative to confronting the structural causes of ecological destruction' (Dryzek 2005: 132). But could individual acts of green consumption be the first and necessary step, even the key to deliver sustainable societies on the basis that 'from little things big things grow'? In other words, can individual changes lead to structural ones? And what about human dignity in general, and that of those living in poverty in particular? How are we going to address the increasingly pressing issue of environmental or climate refugees? And what about the dignity of nature? How are we going to tackle the continuing suffering of animals? With meat being such a major part of our ecological footprints, should we all become vegetarian? Can we envisage animals and nature in general as entities deserving rights, and even citizenship? And, if so, how are we going to balance human rights with animal rights and the rights of nature? And who is going to represent the interests of animals and nature in a potential 'council of all beings'? These are all questions and challenges for us to consider as we continue to experience and explore the intersections between citizenship and the environment into the future.

And, in the meantime: *what is to be done? Or what are we to do?* What does the environment demand from us as citizens? This is an important question, but the answer is far from straightforward. The simple answer is: something. But what exactly? Here, I agree with Zev Trachtenberg, when he argues that 'the most responsible answer is at the same time irreducibly vague: "It depends on the circumstances"' (2010: 349). Once again: context is all. There is no single let alone simple answer. First and foremost, we must recognize that different political activities are appropriate in different contexts. This requires us to examine critically the specific personal, social and political situations in which we operate as citizens, and to judge how to act appropriately, in light of those circumstances, to advance our environmental values. In addition, the fallibility that inevitably accompanies all humans judgements and actions suggests that we might do well to approach environmental citizenship with a sense of irony, that is, recognizing 'the inevitability of failure and error, and at the same time the need to act, with due care, in the very face of that recognition' (Szerszynski 2007: 351). In the end, when all is said and done, the best answer to the question of 'what are we to do?' is always to do 'the best we can'.

Bibliography

Abbey, E. (1975) *The Monkey Wrench Gang*. New York: Avon.

Acosta, A. (2008) 'El Buen Vivir, una oportunidad por construir'. *Ecuador Debate* 75: 33–47.

Adams, S. A. and S. E. Porter (2008) 'Paul The Roman Citizen: Roman Citizenship in the Ancient World and its Importance for Understanding Acts 22:22–29', in S. E. Porter (ed.), *Paul: Jew, Greek, and Roman*. Leiden: Brill, 309–326.

Agrawal, A. (2005) *Environmentality: Technologies of Government and the Making of Subjects*. London: Duke University Press.

Aguilar, A. and N. Popović (1994) 'Lawmaking in the United Nations: The UN Study on Human Rights and the Environment'. *Review of European Community & International Environmental Law* 3(4): 197–205.

Agyeman, J. and B. Evans (2006) 'Justice, Governance, and Sustainability: Perspectives on Environmental Citizenship from North America and Europe', in A. Dobson and D. Bell (eds.), *Environmental Citizenship*. Cambridge, MA: The MIT Press, 185–206.

Alfonsi, A., S. MacGregor and T. Doyle (2014) 'Context Is All: The Many Environmentalisms', in T. Doyle and S. MacGregor (eds.), *Environmental Movements around the World: Shades of Green in Politics and Culture*. Santa Barbara, CA: Praeger, 359–373.

Anderson, A. (1997) *Media, Culture and the Environment*. New Brunswick, NJ: Rutgers University Press.

Anderson, T. L. and D. R. Leal (2001) *Free Market Environmentalism*. Revised edition. New York: Palgrave.

Archibugi, D. (2008) *The Global Commonwealth of Citizens: Toward Cosmopolitan Democracy*. Princeton, NJ: Princeton University Press.

Arendt, H. (1958) *The Human Condition*. Chicago, IL: Chicago University Press.

Arevalo, J. A. (2010) 'Critical Reflective Organizations: An Empirical Observation of Global Active Citizenship and Green Politics'. *Journal of Business Ethics* 96(2): 299–316.

Athanasiou, T. (1996) 'The Age of Greenwashing'. *Capitalism, Nature, Socialism* 7(1): 1–36.

Aubauer, H. (2011) 'Development of Ecological Footprint to an Essential Economic and Political Tool'. *Sustainability* 3(4): 649–665.

Ayres, R. U. (2000) 'Commentary on the Utility of the Ecological Footprint Concept'. *Ecological Economics* 32(3): 347–349.

Bakan, J. (2004) *The Corporation: The Pathological Pursuit of Profit and Power*. Toronto: Penguin.

Baker, L. E. (2005) 'Tending Cultural Landscapes and Food Citizenship in Toronto's Community Gardens'. *Geographical Review* 94(3): 305–325.

Baker, S. (2006) *Sustainable Development*. London: Routledge.

Bakker, K. (2007) 'The "Commons" versus the "Commodity": Alter-globalization, Anti-privatization and the Human Right to Water in the Global South'. *Antipode* 39(3): 430–455.

Banerjee, S. B. (2008) 'Corporate Social Responsibility: The Good, the Bad and the Ugly'. *Critical Sociology* 34(1): 51–79.

Barbalet, J. M. (1988) *Citizenship Rights, Struggle and Class Inequality*. Minneapolis, MN: University of Minnesota Press.

Barker, M. J. (1970) 'The Environmental Citizenship: Where to Begin'. *Art Education* 23(7): 33–35.

Barnett, C. and D. Scott (2007) 'The Reach of Citizenship: Locating the Politics of Industrial Air Pollution in Durban and Beyond'. *Urban Forum* 18(4): 289–309.

Barr, S. (2008) *Environment and Society: Sustainability, Policy and the Citizen*. Aldershot: Ashgate.

Barry, J. (1999) *Rethinking Green Politics*. London: Sage.

Barry, J. (2002) 'Vulnerability and Virtue: Democracy, Dependency, and Ecological Stewardship', in B. Minteer and B. P. Taylor (eds.), *Democracy and the Claims of Nature*. Lanham, MD: Rowman and Littlefield, 133–152.

Barry, J. (2006) 'Resistance Is Fertile: From Environmental to Sustainability Citizenship', in A. Dobson and D. Bell (eds.), *Environmental Citizenship*. Cambridge, MA: MIT Press, 21–48.

Barry, J. (2007) *Environment and Social Theory*. 2nd edition. London: Routledge.

Bauböck, R. (1994) *Transnational Citizenship*. Aldershot: Edward Elgar.

Bautista, L. B. (2007) 'Bioprospecting or Biopiracy: Does the TRIPS Agreement Undermine the Interests of Developing Countries?' *Philippine Law Journal* 82(1): 14–33.

Beck, U. (1999) *World Risk Society*. Cambridge and Malden, MA: Polity Press/ Blackwell Publishers.

Beck, U. (2006) *The Cosmopolitan Vision*. Cambridge: Polity Press.

Beck, U. (2010) 'Climate for Change, or How to Create a Green Modernity?' *Theory, Society & Culture* 27(2–3): 254–266.

Beck, U. (2011) 'Cosmopolitanism as Imagined Communities of Global Risk'. *American Behavioral Scientist* 55(10): 1346–1361.

Beder, S. (2002) *Global Spin: The Corporate Assault on Environmentalism*. Revised Edition. Carlton North, Vic.: Scribe Publications.

Bell, D. (2004) 'Creating Green Citizens? Political Liberalism and Environmental Education'. *Journal of Philosophy of Education* 38(1): 37–53.

Bell, D. (2005) 'Liberal Environmental Citizenship'. *Environmental Politics* 14(2): 179–194.

Bellamy, R. (2008) *Citizenship: A Very Short Introduction*. Oxford: Oxford University Press.

Best, S. (1998) 'Murray Bookchin's Theory of Social Ecology: An Appraisal of *The Ecology of Freedom*'. *Organization and Environment* 11(3): 334–353.

Bevir, M. (2010) *Democratic Governance*. Princeton, NJ: Princeton University Press.

Bevir, M. (2011) 'Governance and Governmentality after Neoliberalism'. *Policy and Politics* 39(4): 457–471.

Bhandari, B. B. and O. Abe (2001) 'Corporate Citizenship and Environmental Education'. *International Review for Environmental Strategies* 2(1): 61–77.

Birke, L. (2000) 'Supporting the Underdog: Feminism, Animal Rights and Citizenship in the Work of Alice Morgan Wright and Edith Goode'. *Women's History Review* 9(4): 693–719.

Blühdorn, I. (1997) 'A Theory of Post-Ecologist Politics'. *Environmental Politics* 6(3): 125–147.

Blühdorn, I. (2000) *Post-ecologist Politics: Social Theory and the Abdication of the Ecologist Paradigm*. London: Routledge.

Blühdorn, I. (2007) 'Sustaining the Unsustainable: Symbolic Politics and the Politics of Simulation'. *Environmental Politics* 16(2): 251–275.

Blum, E. (2008) *Love Canal Revisited: Race, Class, and Gender in Environmental Activism*. Lawrence, KS: University Press of Kansas.

Bocarejo, D. (2009) 'Deceptive Utopias: Violence, Environmentalism and the Regulation of Multiculturalism in Colombia'. *Law and Policy* 31(3): 307–329.

Bohman, J. (2004) 'Republican Cosmopolitanism'. *Journal of Political Philosophy* 12(3): 336–353.

Bourke, L., D. Bamber and M. Lyons (2012) 'Global Citizens: Who Are They?' *Education, Citizenship and Social Justice* 7(2): 161–174.

Bowden, B. (2003) 'The Perils of Global Citizenship'. *Citizenship Studies* 7(3): 349–362.

Bookchin, M. (1995) *From Urbanization to Cities: Toward a New Politics of Citizenship*. London: Cassell.

Brand, P. (2007) 'Green Subjection: The Politics of Neoliberal Urban Environmental Management'. *International Journal of Urban and Regional Research* 31(1): 616–632.

Brandl, E. and H. Bungert (1992) 'Constitutional Entrenchment of Environmental Protection: A Comparative Analysis of Experiences Abroad'. *Harvard Environmental Law Review* 16(1): 1–100.

Brockington, D. (2008) 'Powerful Environmentalisms: Conservation, Celebrity and Capitalism'. *Media, Culture & Society* 30(4): 551–568.

Brodie, J. (2004) 'Introduction: Globalization and Citizenship beyond the National State'. *Citizenship Studies* 8(4): 323–332.

Brook, A. (2011) 'Ecological Footprint Feedback: Motivating or Discouraging?' *Social Influence* 6(2): 113–128.

Brown, W. (2005) *Edgework: Critical Essays on Knowledge and Politics*. Princeton, NJ: Princeton University Press.

Brubaker, R. (1992) *Citizenship and Nationhood in France and Germany*. Cambridge, MA: Harvard University Press.

Buckingham, S. and R. Kulcur (2009), 'Gendered Geographies of Environmental Injustice'. *Antipode* 41(4): 659–683.

Buckingham-Hatfield, S. (2000) *Gender and Environment*. London and New York: Routledge.

Bullard, R. D. (1990) *Dumping in Dixie: Race, Class, and Environmental Quality*. Boulder, CO: Westview Press.

Bullen, A. and M. Whitehead (2005) 'Negotiating the Networks of Space, Time and Substance: A Geographical Perspective on the Sustainable Citizen'. *Citizenship Studies* 9(5): 499–516.

Burchell, D. (2002) 'Ancient Citizenship and its Inheritors', in E. F. Isin and B. S. Turner (eds.), *Handbook of Citizenship Studies*. London: Sage, 88–104.

Burke, R. (2010) *Decolonization and the Evolution of International Human Rights*. Philadelphia, PA: University of Pennsylvania Press.

Butler, C. (2010) 'Morality and Climate Change: Is Leaving Your TV on Standby a Risky Behaviour?' *Environmental Values* 19(2): 169–192.

Butler, J. (1990) *Gender Trouble: Feminism and the Subversion of Identity*. New York: Routledge.

Butler, J. (1994) 'Gender as Performance: An Interview with Judith Butler'. *Radical Philosophy: A Journal of Socialist and Feminist Philosophy* 67: 32–39.

Buttel, F. H. (2000) 'Ecological Modernization as Social Theory'. *Geoforum* 31(1): 57–65.

Buttel, F. H. (2005) 'The Environmental and Post-Environmental Politics of Genetically Modified Crops and Foods'. *Environmental Politics* 14(3): 309–323.

Cairncross, F. (1995) *Green, Inc.: Guide to Business and the Environment*. London: Earthscan.

Calhoun, C. (2002) 'The Class Consciousness of Frequent Travelers: Toward a Critique of Actually Existing Cosmopolitanism'. *The South Atlantic Quarterly* 101(4): 869–897.

Cannavò, P. F. (2012) 'Ecological Citizenship, Time, and Corruption: Aldo Leopold's Green Republicanism'. *Environmental Politics* 21(6): 864–881.

Cao, B. (2014) 'Dam-Nation: The Struggle over the Future of the Brazilian Amazon', in T. Doyle and S. MacGregor (eds.), *Environmental Movements around the World: Shades of Green in Politics and Culture*. Santa Barbara, CA: Praeger, 49–75.

Carlsson, M. and B. B. Jensen (2006) 'Encouraging Environmental Citizenship: The Roles and Challenges for Schools', in A. Dobson and D. Bell (eds.), *Environmental Citizenship*. Cambridge, MA: The MIT Press, 237–261.

Carson, R. (1962) *Silent Spring*. Boston, MA: Houghton Mifflin.

Carter, C. and S. L. Kravits (2013) *Keys to Success: Cultural Awareness and Global Citizenship*. Boston, MA: Pearson.

Carter, N. (2007) *The Politics of the Environment: Ideas, Activism, Policy*. 2nd edition. New York: Cambridge University Press.

Carter, N. and M. Huby (2005) 'Ecological Citizenship and Ethical Investment'. *Environmental Politics* 14(2): 255–272.

Castles, S. and A. Davidson (2000) *Citizenship and Migration: Globalization and the Politics of Belonging*. New York: Routledge.

Cavalieri, P. and P. Singer (1993a) 'Preface', in P. Cavalieri and P. Singer (eds.), *The Great Ape Project: Equality beyond Humanity*. London: Fourth Estate Limited, 1–3.

Cavalieri, P. and P. Singer (1993b) 'The Great Ape Project – and Beyond', in P. Cavalieri and P. Singer (eds.), *The Great Ape Project: Equality beyond Humanity*. London: Fourth Estate Limited, 304–312.

Cavalieri, P. and P. Singer (eds.) (1993c) *The Great Ape Project: Equality beyond Humanity*. London: Fourth Estate Limited.

Cederlöf, G. and M. Rangarajan (2009) 'Predicaments of Power and Nature in India: An Introduction'. *Conservation and Society* 7(4): 221–226.

Chamberlain, D. M. (2007) 'A Path Forward or a Dead End? The Western Political Tradition and the Struggle to Accommodate Green Conceptions of Citizenship'. *Environmental Practice* 9(2): 136–139.

Chapman, G., K. Kumar, C. Fraser and I. Gaber (1997) *Environmentalism and the Mass Media: The North/South Divide*. New York: Routledge.

Christoff, P. (2000) 'Environmental Citizenship', in W. Hudson and J. Kane (eds.), *Rethinking Australian Citizenship*. Cambridge: Cambridge University Press, 200–214.

Cintra, Y. C. and D. Carter (2012) 'Internalising Sustainability: Reflections on Management Control in Brazil'. *International Journal of Strategic Management* 12(2): 108–125.

Clark, N. and N. Stevenson (2003) 'Care in the Time of Catastrophe: Citizenship, Community and the Ecological Imagination'. *Journal of Human Rights* 2(2): 235–246.

Clarke, T. (1999) 'Twilight of the Corporation'. *The Ecologist* 29(2): 158–161.

Connelly, J. (2006) 'The Virtues of Environmental Citizenship', in A. Dobson and D. Bell (eds.), *Environmental Citizenship*. Cambridge, MA: The MIT Press, 49–73.

Connor, S. (2009) 'Forty Years Since the First Picture of Earth from Space'. *The Independent*, 10 January 2009. Available at: www.independent.co.uk/news/science/forty-years-since-the-first-picture-of-earth-from-space-1297569.html (Accessed 22/06/13).

Coppolecchia, E. (2010) 'The Greenwashing Deluge: Who Will Rise above the Waters of Deceptive Advertising?' *University of Miami Law Review* 6(4): 1353–1406.

Cordero, E. C., A. M. Todd and D. Abellera (2008) 'Climate Change Education and the Ecological Footprint'. *Bulletin of the American Meteorological Society* 89(6): 865–872.

Crane, A., D. Matten and J. Moon (2008a) *Corporations and Citizenship*. Cambridge: Cambridge University Press.

Crane, A., D. Matten and J. Moon (2008b) 'Ecological Citizenship and the Corporation: Politicizing the New Corporate Environmentalism'. *Organization and Environment* 21(4): 371–389.

Cronin, A. (2000) *Advertising and Consumer Citizenship: Gender, Images and Rights*. London: Routledge.

Cubitt, S. (2005a) *Nature in Motion Pictures: Mass Media and the Environment*. Amsterdam: Rodopi.

Cubitt, S. (2005b) *EcoMedia*. Amsterdam: Rodopi.

Cudworth, E. (2003) *Environment and Society*. London: Routledge.

Cudworth, E. and S. Hobden (2013) 'Complexity, Ecologism and Posthuman Politics'. *Review of International Studies* 39(3): 643–664.

Curry, P. (2000) 'Redefining Community: Towards an Ecological Republicanism'. *Biodiversity and Conservation* 9: 1059–1071.

Curtin, D. (2002) 'Ecological Citizenship', in E. F. Isin and B. S. Turner (eds.), *Handbook of Citizenship Studies*. London: Sage, 293–305.

Cutting, R., L. Cahoon, J. Flood, L. Horton and M. Schramm (2011) 'Spill the Beans: Good Guide, Walmart and EPA Use Information as Efficient, Market-Based Environmental Regulation'. *Tulane Environmental Law Journal* 24: 291–334.

Dagger, R. (2002) 'Republican Citizenship', in E. F. Isin and B. S. Turner (eds.), *Handbook of Citizenship Studies*. London: Sage, 145–157.

Darier, E. (1996) 'Environmental Governmentality: The Case of Canada's Green Plan'. *Environmental Politics* 5(4): 585–606.

David, R. A. (1991) *The Greening of Business*. Brookfield, VT: Gower in Association with *Business Magazine*.

Davies, J. S. (2012) 'Network Governance Theory: A Gramscian Critique'. *Environment and Planning A* 44(11): 2687–2704.

Dean, H. (2001) 'Green Citizenship'. *Social Policy & Administration* 35(5): 490–505.

Delanty, G. (2000) *Citizenship in a Global Age: Society, Culture, Politics*. Buckingham: Open University Press.

DeLuca, K. M. (1999) *Image Politics: The New Rhetoric of Environmental Activism*. New York: Guilford Press.

Dennis, E. E. and C. L. LaMay (eds.) (1991) *Media and the Environment*. Washington, DC: Island Press.

De Steiguer, J. E. (2006) *The Origins of Modern Environmental Thought*. Tucson, AZ: The University of Arizona Press.

De Tavernier, J. (2012) 'Food Citizenship: Is There a Duty for Responsible Consumption?' *Journal of Agricultural and Environmental Ethics* 25(6): 893–908.

Deumling, D., M. Wackernagel and C. Monfreda (2003) *Eating up the Earth: How Sustainable Food Systems Shrink Our Ecological Footprint*. Oakland, CA: Redefining Progress.

Di Chiro, G. (2006) 'Teaching Urban Ecology: Environmental Studies and the Pedagogy of Intersectionality'. *Feminist Teacher* 16(2): 98–109.

Di Chiro, G. (2008) 'Living Environmentalisms: Coalition Politics, Social Reproduction, and Environmental Justice'. *Environmental Politics* 17(2): 276–298.

Di Chiro, G. (2011) 'Acting Globally: Cultivating a Thousand Community Solutions for Climate Justice'. *Development* 54(2): 232–236.

Dickens, P. (2004) *Society and Nature: Changing Our Environment, Changing Ourselves*. Cambridge: Polity Press.

Dickinson, G. (2005) 'Selling Democracy: Consumer Culture and Citizenship in the Wake of September 11'. *Southern Communication Journal* 70(4): 271–284.

Dickinson, R. and M. Carsky (2005) 'The Consumer as Economic Voter', in R. Harrison, T. Newholm and D. Shaw (eds.), *The Ethical Consumer*. Thousand Oaks, CA: Sage, 25–36.

Dietz, M. G. (1987) 'Context Is All: Feminism and Theories of Citizenship'. *Daedalus* 116(4): 1–24.

Dixon, J. and Y. Frolova (2011) 'Existential Poverty: Welfare Dependency, Learned Helplessness and Psychological Capital'. *Poverty & Public Policy* 3(2): 1–20.

Dobson, A. (2000) 'Ecological Citizenship: A Disruptive Influence?', in C. Pierson and S. Torney (eds.), *Politics at the Edge: The PSA Yearbook 1999*. London: Palgrave Macmillan, 40–62.

Dobson, A. (2003) *Citizenship and the Environment*. Oxford: Oxford University Press.

Dobson, A. (2005) 'Globalisation, Cosmopolitanism and the Environment'. *International Relations* 19(3): 259–273.

Dobson, A. (2006a) 'Thick Cosmopolitanism'. *Political Studies* 54(1): 165–184.

Dobson, A. (2006b) 'Ecological Citizenship: A Defence'. *Environmental Politics* 15(3): 447–451.

Dobson, A. (2007) *Green Political Thought*. 4th edition. London: Routledge.

Dobson, A. (2011) 'Sustainability Citizenship'. Greenhouse Thinktank. Available at: www.sustainablecitizenship.com/about.html (Accessed 10/04/12).

Dobson, A. (2012) 'Ecological Citizenship Revisited', in P. Dauvergne (ed.), *Handbook of Global Environmental Politics*. 2nd edition. Cheltenham: Edward Elgar, 520–529.

Dobson, A. and D. Bell (eds.) (2006) *Environmental Citizenship*. Cambridge, MA: The MIT Press.

Dobson, A. and R. Eckersley (eds.) (2006) *Political Theory and the Ecological Challenge*. Cambridge: Cambridge University Press.

Doherty, B. (2002) *Ideas and Action in the Green Movement*. London: Routledge.

Doherty, B. and M. de Geus (1996) *Democracy and Green Political Thought: Sustainability, Rights and Citizenship*. London: Routledge.

Doherty, B. and T. Doyle (2006) 'Beyond Borders: Transnational Politics, Social Movements and Modern Environmentalisms'. *Environmental Politics* 15(5): 697–712.

Donaldson, S. and W. Kymlicka (2011) *Zoopolis: A Political Theory of Animal Rights*. Oxford: Oxford University Press.

Dower, N. and J. Williams (eds.) (2002) *Global Citizenship: A Critical Introduction*. New York: Routledge.

Dowie, M. (2009) *Conservation Refugees: The Hundred-Year Conflict between Global Conservation and Native Peoples*. Cambridge, MA: The MIT Press.

Doyle, T. (2005) *Environmental Movements in Minority and Majority Worlds: A Global Perspective*. New Brunswick, NJ: Rutgers University Press.

Doyle, T. and D. McEachern (2008) *Environment and Politics*. 3rd edition. London: Routledge.

Doyle, T. and A. Simpson (2006) 'Traversing More than Speed Bumps: Green Politics under Authoritarian Regimes in Burma and Iran'. *Environmental Politics* 15(5): 750–767.

Driessens, O. (2012) 'The Celebritization of Society and Culture: Understanding the Structural Dynamics of Celebrity Culture'. *International Journal of Cultural Studies* 16(6): 641–657.

Dryzek, J. S. (1997) *The Politics of the Earth: Environmental Discourses*. New York: Oxford University Press.

Dryzek, J. S. (2005) *The Politics of the Earth: Environmental Discourses*. 2nd edition. New York: Oxford University Press.

Dryzek, J. S. and D. Schlosberg (eds.) (2005) *Debating the Earth: The Environmental Politics Reader*. 2nd edition. Oxford: Oxford University Press.

Duncan, D. and K. Burns (2009) *The National Parks: America's Best Idea*. New York: Alfred A. Knopf.

Dunlap, T. R. (ed.) (2008) *DDT, Silent Spring, and the Rise of Environmentalism*. Seattle, WA: University of Washington Press.

Easterbrook, G. (1995) *A Moment on the Earth: The Coming Age of Environmental Optimism*. New York: Viking.

Ebbesson, J. (2011) 'A Modest Contribution to Environmental Democracy and Justice in Transboundary Contexts: The Combined Impact of the Espoo Convention and Aarhus Convention'. *Review of European Community and International Environmental Law* 20(3): 248–257.

Eckersley, R. (2004) *The Green State: Rethinking Democracy and Sovereignty*. Cambridge, MA: The MIT Press.

Edwards, T. (2000) *Contradictions of Consumption: Concepts, Practices and Politics in Consumer Society*. Buckingham: Open University Press.

Ehrlich, P. R. (1968) *The Population Bomb*. New York: Ballantine Books.

Elliot, D. (2003) *Energy, Society and Environment: Technology for a Sustainable Future*. 2nd edition. London: Routledge.

Ellis, R. and C. Waterton (2004), 'Environmental Citizenship in the Making: The Participation of Volunteer Naturalists in UK Biological Recording and Biodiversity Policy'. *Science and Public Policy* 31(2): 95–105.

Ellison, N. (1997) 'Towards a New Social Politics: Citizenship and Reflexivity in Late Modernity'. *Sociology* 31(4): 697–717.

Elshtain, J. B. (1981) *Public Man, Private Woman*. Oxford: Martin Robertson.

Enroth, H. (2014) 'Governance: The Art of Governing after Governmentality'. *European Journal of Social Theory* 17(1): 60–76.

Enslin, P. and P. White (2003) 'Democratic Citizenship', in N. Blake, P. Smeyers, R. D. Smith and P. Standish (eds.), *The Blackwell Guide to the Philosophy of Education*. Oxford: Blackwell Publishing, 110–125.

Environment Canada (1993) 'From the Mountains to the Sea: A Journey in Environmental Citizenship'. Canada: Published by authority of the Minister of the Environment. Available at: http://infohouse.p2ric.org/ref/14/13416.pdf (Accessed 23/07/13).

Estes, R. W. (1996) *Tyranny of the Bottom Line: Why Corporations Make Good People Do Bad Things*. San Francisco, CA: Berrett-Koehler.

Esteva, G. and M. S. Prakash (1994) 'Editorial: From Global to Local Thinking'. *The Ecologist* 24(5): 162–163.

European Commission (2006) Climate Change Campaign 'You Control Climate Change'. Available at: http://europa.eu/rapid/press-release_MEMO-06-218_en.htm (Accessed 30/01/13).

European Commission (2007) 'Change: Turn Down. Switch Off. Recycle. Walk. Make a Pledge!' Luxembourg: Office for Official Publications of the European Communities. Available at: http://ec.europa.eu/clima/sites/campaign/pdf/toolkit_en.pdf (Accessed 23/07/13).

Evans, J. P. (2012) *Environmental Governance*. London: Routledge.

Faber, D. (2005) 'Building a Transnational Environmental Justice Movement: Obstacles and Opportunities in the Age of Globalization', in J. Bandy and J. Smith (eds.), *Coalitions Across Borders: Transnational Protest and the Neoliberal Order*. Oxford: Rowman & Littlefield, 45–80.

Fadaee, S. (2011) 'Environmental Movements in Iran: Application of the New Social Movement Theory in the Non-European Context'. *Social Change* 41(1): 79–96.

Fagan, A. (2004) *Environment and Democracy in the Czech Republic: The Environmental Movement in the Transition Process*. Cheltenham: Edward Elgar.

Fagan, A. (2006) 'Neither "North" nor "South": The Environment and Civil Society in Post-Conflict Bosnia-Herzegovina'. *Environmental Politics* 15(5): 787–802.

Falk, R. (1994) 'The Making of Global Citizenship', in B. van Steenbergen (ed.), *The Condition of Citizenship*. London: Sage, 127–140.

Fan, M. (2008) 'Environmental Citizenship and Sustainable Development: The Case of Waste Facility Siting in Taiwan'. *Sustainable Development* 16(6): 311–389.

Feeney, L. (2013) 'Citizens, Not Consumers, Are Key to Solving Climate Crisis'. *Moyers & Company*, 14 June 2013. Available at: http://billmoyers.com/2013/01/04/citizens-not-consumers-are-key-to-solving-climate-crisis/ (Accessed 18/02/14).

Feichtinger, J. and M. Pregernig (2005) 'Imagined Citizens and Participation: Local Agenda 21 in Two Communities in Sweden and Austria'. *Local Environment* 10(3): 229–242.

Firth, C., D. Maye and D. Pearson (2011) 'Developing "Community" in Community Gardens'. *Local Environment* 16(6): 555–568.

Fishman, C. (2006) *The Wal-Mart Effect: How the World's Most Powerful Company Really Works – and How It's Transforming the American Economy*. New York: Penguin.

Fletcher, S. and J. Potts (2007) 'Ocean Citizenship: An Emergent Geographical Concept'. *Coastal Management* 35: 511–524.

Fletcher, T. (2002) 'Neighbourhood Change at Love Canal: Contamination, Evacuation and Resettlement'. *Land Use Policy* 19: 311–323.

Foucault, M. (1991) *Discipline and Punish: The Birth of the Prison*. London: Penguin.

Franz, J. and E. Papyrakis (2011) 'Online Calculators of Ecological Footprint: Do They Promote or Dissuade Sustainable Behaviour?' *Sustainable Development* 19(6): 391–401

Freeland, C. (2012) *Plutocrats: The Rise of the New Global Super-Rich*. London: Penguin.

Frey, S. B. (2003) 'Flexible Citizenship for a Global Society'. *Politics, Philosophy and Economics* 2(1): 93–114.

Friedman, J. (2002) *The Prospect of Cities*. Minneapolis, MN: University of Minnesota Press.

Friedman, M. (1970) 'The Social Responsibility of Business Is to Increase Its Profits'. *The New York Times Magazine*, 13 September 1970. Available at: www.umich.edu/~thecore/doc/Friedman.pdf (Accessed 24/02/13).

Fuentes-George, K. (2013) 'Neoliberalism, Environmental Justice, and the Convention on Biological Diversity: How Problematizing the Commodification of Nature Affects Regime Effectiveness'. *Global Environmental Politics* 13(4): 144–163.

Gaard, G. (2010) 'New Directions for Ecofeminism: Toward a More Feminist Ecocriticism'. *Interdisciplinary Studies in Literature and Environment* 17(4): 643–665.

Gaard, G. (2011) 'Ecofeminism Revisited: Rejecting Essentialism and Re-Placing Species in a Material Feminist Environmentalism'. *Feminist Formations* 23(2): 26–53.

Gabrielson, T. (2008) 'Green Citizenship: A Review and Critique'. *Citizenship Studies* 12(4): 429–446.

Gabrielson, T. and K. Parady (2010) 'Corporeal Citizenship: Rethinking Green Citizenship through the Body'. *Environmental Politics* 19(3): 374–391.

Gebbels, S., S. M. Evans and J. E. Delany (2011) 'Promoting Environmental Citizenship and Corporate Social Responsibility Through School/Industry/ University Partnership'. *Journal of Biological Education* 45(1): 13–19.

Gibbs, D. (2000) 'Ecological Modernization, Regional Economic Development and Regional Development Agencies'. *Geoforum* 31(1): 9–19.

Gibbs, L. (2002) 'Citizen Activism for Environmental Health: The Growth of a Powerful New Grassroots Health Movement'. *The Annals of the American Academy of Political and Social Science (AAPSS)* 584: 97–109.

Gibbs, L. (2011) *Love Canal and the Birth of the Environmental Health Movement*. Updated edition. Washington, DC: Island Press.

Gillespie, E. (2008) 'Stemming the Tide of "Greenwash"'. *Consumer Policy Review* 18(3): 79–83.

Gilson, D. (2005) '"I Will Disappear into the Forest": An Interview with Wangari Maathai'. *Mother Jones*, 5 January 2005. Available at: www.motherjones. com/news/qa/2005/01/wangari_maathai.html (Accessed 12/12/13).

Glickman, L. B. (2009) *Buying Power: A History of Consumer Activism in America*. Chicago, IL: The University of Chicago Press.

Godrej, F. (2012) 'Ascetics, Warriors, and a Gandhian Ecological Citizenship'. *Political Theory* 40(4): 437–465.

Golay, C. and I. Biglino (2013) 'Human Rights Responses to Land Grabbing: A Right to Food Perspective'. *Third World Quarterly* 34(9): 1630–1650.

Gold, J. R. and G. Revill (2004) *Representing the Environment*. London: Routledge.

Goodell, J. (2010) *How to Cool the Planet: Geoengineering and the Audacious Quest to Fix Earth's Climate*. Boston, MA: Houghton Mifflin Harcourt.

Goodin, R. E., C. Pateman and R. Pateman (1997) 'Simian Sovereignty'. *Political Theory* 25(6): 821–849.

Goodman, M. K. and J. Littler (2013) 'Celebrity Ecologies: Introduction'. *Celebrity Studies* 4(3): 269–275.

Gorman, D. (2006) *Imperial Citizenship: Empire and the Question of Belonging*. Manchester: Manchester University Press.

Gottlieb, D., E. Vigoda-Gadot, A. Haim and M. Kissinger (2012). 'The Ecological Footprint as an Educational Tool for Sustainability: A Case Study Analysis in an Israeli Public High School'. *International Journal of Educational Development* 32(3): 193–200.

Gough, S. and W. Scott (2006) 'Promoting Environmental Citizenship through Learning: Toward a Theory of Change', in A. Dobson and D. Bell (eds.), *Environmental Citizenship*. Cambridge, MA: The MIT Press, 263–285.

Grimm, D. (2014) *Citizen Canine: Our Evolving Relationship with Cats and Dogs*. New York: PublicAffairs.

Gudynas, E. (2011a) 'Buen Vivir: Today's Tomorrow'. *Development* 54(4): 441–447.

Gudynas, E. (2011b) 'Desarrollo, Derechos de la Naturaleza y Buen Vivir después de Montecristi', in G. Weber (ed.), *Debates sobre cooperación y modelos de desarrollo. Perspectivas desde la sociedad civil en el Ecuador*. Quito: Centro de Investigaciones CIUDAD y Observatorio de la Cooperación al Desarrollo, 83–102.

Gudynas, E. (2011c) 'Los derechos de la Naturaleza en serio: Respuestas y aportes desde la ecología política', in A. Acosta and E. Martínez (eds.), *La Naturaleza con Derechos: De la filosofía a la política*. Quito: AbyaYala y Universidad Politécnica Salesiana, 239–258.

Guha, R. (1997a) 'The Authoritarian Biologist and the Arrogance of Anti-Humanism: Wildlife Conservation in the Third World'. *The Ecologist* 27(1): 14–21.

Guha, R. (1997b) 'Radical American Environmentalism and Wilderness Preservation'. *Environmental Ethics* 11(1): 71–83.

Guha, R. and J. Martínez-Alier (1997) *Varieties of Environmentalism: Essays North and South*. London: Earthscan Publications.

Hailwood, S. (2005) 'Environmental Citizenship as Reasonable Citizenship'. *Environmental Politics* 14(2): 195–210.

Hall, M. (2011) 'Beyond the Human: Extending Ecological Anarchism'. *Environmental Politics* 20(3): 374–390.

Halweil, B. (2005) 'The Rise of Food Democracy'. *UN Chronicle* 42(1): 71–73.

Hanasz, W. (2006) 'Toward Global Republican Citizenship?' *Social Philosophy and Policy* 23(1): 282–302.

Hancock, J. (2003) *Environmental Human Rights: Power, Ethics and Law*. Aldershot: Ashgate.

Hansen, A. (2010) *Environment, Media and Communication*. London: Routledge.

Haq, G. and A. Paul (2012) *Environmentalism since 1945*. London: Routledge.

Hardin, G. (1968) 'The Tragedy of the Commons'. *Science* 163: 1243–1248.

Hargrove, E. (2001) 'Environmental Citizenship'. Available at: www.cep.unt.edu/citizen.htm (Accessed 13/12/11).

Harris, L. M. (2011) 'Neo(liberal) Citizens of Europe: Politics, Scales, and Visibilities of Environmental Citizenship in Contemporary Turkey'. *Citizenship Studies* 15(6–7): 837–859.

Harvey, D. (2005) *A Brief History of Neoliberalism*. Oxford: Oxford University Press.

Hay, A. M. (2009) 'Recipe for Disaster: Motherhood and Citizenship at Love Canal'. *Journal of Women's History* 21(1): 111–134.

Hayek, F. A. (1944) *The Road to Serfdom*. Chicago, IL: University of Chicago Press.

Hayward, B. (2012) *Children, Citizenship and Environment: Nurturing a Democratic Imagination in a Changing World*. London: Routledge

Hayward, T. (2000) 'Constitutional Environmental Rights: A Case for Political Analysis'. *Political Studies* 48(3): 558–572.

Hayward, T. (2005) *Constitutional Environmental Rights*. Oxford: Oxford University Press.

Hayward, T. (2006) 'Ecological Citizenship: Justice, Rights and the Virtue of Resourcefulness'. *Environmental Politics* 15(3): 435–446.

Heater, D. (1999) *What Is Citizenship?* Cambridge: Polity Press.

Heater, D. (2002) *World Citizenship: Cosmopolitan Thinking and its Opponents*. London: Continuum.

Heater, D. (2004a) *A Brief History of Citizenship*. Edinburgh: Edinburgh University Press.

Heater, D. (2004b) *Citizenship: The Civic Ideal in World History, Politics and Education*. Manchester: Manchester University Press.

Hecht, S. B. (2011) 'The New Amazon Geographies: Insurgent Citizenship, "Amazon Nation" and the Politics of Environmentalisms'. *Journal of Cultural Geography* 28(1): 203–223.

Heise, U. (2008) *Sense of Place and Sense of Planet: The Environmental Imagination of the Global*. Oxford: Oxford University Press.

Held, D. (2004) *Global Covenant: The Social Democratic Alternative to the Washington Consensus*. Malden, MA: Polity.

Henriques, A. (2005) 'Corporations: Amoral Machines or Moral Persons?' *Business and Professional Ethics Journal* 24(3): 91–98.

Hickman, L. (2010) 'James Lovelock: Humans Are Too Stupid to Prevent Climate Change'. *The Guardian*, 29 March 2010. Available at: www.theguardian.com/science/2010/mar/29/james-lovelock-climate-change (Accessed 08/08/13).

Hindess, B. (2002) 'Neo-liberal Citizenship'. *Citizenship Studies* 6(2): 127–143.

Hiskes, R. (2005) 'The Rights to a Green Future: Human Rights, Environmentalism, and Intergenerational Justice'. *Human Rights Quarterly* 27: 1346–1364.

Hobson, K. (2004) 'Sustainable Consumption in the United Kingdom: The "Responsible" Consumer and Government at "Arm's Length"'. *Journal of Environment & Development* 13(2): 121–139.

Hochschild, A. R. (2012) *The Outsourced Self: What Happens When We Pay Others to Live Our Lives for Us*. New York: Metropolitan Books.

Hochstetler, K. and M. E. Keck (2007) *Greening Brazil: Environmental Activism in State and Society*. Durham, NC: Duke University Press.

Holden, A. (2008) *Environment and Tourism*. 2nd edition. London: Routledge.

Holston, J. (1998) 'Spaces of Insurgent Citizenship', in L. Sandercock (ed.), *Making the Invisible Visible: A Multicultural Planning History*. Berkeley, CA: University of California Press, 37–56.

Holston, J. (2007) *Insurgent Citizenship: Disjunctions of Democracy and Modernity in Brazil*. Princeton, NJ: Princeton University Press.

Houser, N. O. (2005) 'Inquiry Island: Social Responsibility and Ecological Sustainability in the Twenty-first Century'. *Social Studies* 96(3): 127–132.

Howes, M. (2005) *Politics and the Environment: Risk and the Role of Government and Industry*. London: Earthscan.

Huggan, G. (2013) *Nature's Saviours: Celebrity Conservationists in the Television Age*. London: Routledge.

Hulme, M. and M. Mahony (2012) 'Climate Change: What Do We Know about the IPCC?' *Progress in Physical Geography* 34(5): 705–718.

Humphreys, D. (2009) 'Environmental and Ecological Citizenship in Civil Society'. *The International Spectator* 44(1): 171–183.

Inglehart, R. (1977) *The Silent Revolution: Changing Values and Political Styles Among Western Publics*. Princeton, NJ: Princeton University Press.

Inglehart, R. (1990) *Culture Shift in Advanced Industrial Society*. Princeton, NJ: Princeton University Press.

Inglehart, R. (2008) 'Changing Values among Western Publics from 1970 to 2006'. *West European Politics* 31(1–2): 130–146.

Ingram, D. (2000) *Green Screen: Environmentalism and Hollywood Cinema*. Exeter: University of Exeter Press.

Insch, A. (2008) 'Online Communication of Corporate Environmental Citizenship: A Study of New Zealand's Electricity and Gas Retailers'. *Journal of Marketing Communications* 14(2): 139–153.

Inthorn, S. and M. Reder (2011) 'Discourses of Environmental Citizenship: How Television Teaches Us to Be Green'. *International Journal of Media and Cultural Politics* 7(1): 37–54.

Isin, E. F. and B. S. Turner (2002a) 'Citizenship Studies: An Introduction', in E. F. Isin and B. S. Turner (eds.), *Handbook of Citizenship Studies*. London: Sage, 1–10.

Isin, E. F. and B. S. Turner (eds.) (2002b) *Handbook of Citizenship Studies*. London: Sage.

Jagers, S. (2009) 'In Search of the Ecological Citizen'. *Environmental Politics* 18(1): 18–36.

Jelin, E. (2000) 'Towards a Global Environmental Citizenship?' *Citizenship Studies* 4(1): 47–63.

Jickling, B. (2003) 'Environmental Education and Environmental Advocacy: Revisited'. *The Journal of Environmental Education* 34(2): 20–27.

Johnston, J. (2008) 'The Citizen-Consumer Hybrid: Ideological Tensions and the Case of Whole Foods Market'. *Theory and Society* 37(3): 229–270.

Joosse, P. (2012) 'Elves, Environmentalism, and "Eco-Terror": Leaderless Resistance and Media Coverage of the Earth Liberation Front'. *Crime Media Culture* 8(1): 75–93.

Joppke, C. (2007) 'Transformation of Citizenship: Status, Rights, Identity'. *Citizenship Studies* 11(1): 37–48.

Jubas, K. (2007) 'Conceptual Con/fusion in Democratic Societies: Understandings and Limitations of Consumer-Citizenship'. *Journal of Consumer Culture* 7(2): 231–254.

Kääpä, P. and T. Gustafsson (eds.) (2013) *Transnational Ecocinema: Film Culture in an Era of Ecological Transformation*. Bristol: Intellect.

Karliner, J. (1997) *The Corporate Planet: Ecology and Politics in the Age of Globalization*. San Francisco, CA: Sierra Club Books.

Kashmanian, R. M., R. P. Wells and C. Keenan (2011) 'Corporate Environmental Sustainability Strategy: Key Elements'. *Journal of Corporate Citizenship* 44: 107–130.

Kaufman, L. (2010) 'At 40, Earth Day Is Now Big Business'. *The New York Times*, 21 April 2010. Available at: www.nytimes.com/2010/04/22/business/energy-environment/22earth.html?_r=0 (Accessed 24/03/13).

Keiner, M. (2006) 'Rethinking Sustainability: Editor's Introduction', in M. Keiner (ed.), *The Future of Sustainability*. Dordrecht: Springer, 1–15.

Ki-moon, B. (2012a) 'A Global Movement for Change'. *The Express Tribune*, 14 June 2012. Available at: www.un.org/sg/articles/articleFull.asp?TID=129&Type=Op-Ed&h=0 (Accessed 30/01/14).

Ki-moon, B. (2012b) 'Ban Ki-moon: The Momentum for Change at Rio+20 Is Irreversible'. *The Guardian*, 18 June 2012. Available at: www.theguardian.com/environment/2012/jun/18/rio-20-ban-ki-moon (Accessed 30/01/14).

King, D. L. (1994) 'Captain Planet and the Planeteers: Kids, Environmental Crisis, and Competing Narratives of the New World Order'. *Sociological Quarterly* 35(1): 103–120.

Kintisch, E. (2010) *Hack the Planet: Science's Best Hope – or Worst Nightmare – for Averting Climate Catastrophe*. Hoboken, NJ: John Wiley & Sons.

Kitzes, J. and M. Wackernagel (2009) 'Answers to Common Questions in Ecological Footprint Accounting'. *Ecological Indicators* 9(4): 812–817.

Kitzes, J., M. Wackernagel, J. Loh, A. Peller, S. Goldfinger, D. Cheng and K. Tea (2008) 'Shrink and Share: Humanity's Present and Future Ecological Footprint'. *Philosophical Transactions of the Royal Society* 363: 467–475.

Klijn, E.-H. and C. Skelcher (2007) 'Democracy and Governance Networks: Compatible or Not?' *Public Administration* 85(3): 587–608.

Koerner, A. (2013) 'Oscars 2013 Red Carpet – Helen Hunt in Eco-Friendly H&M'. *Ecorazzi.Com*, 24 February 2013. Available at: www.ecorazzi.com/2013/02/24/oscars-2013-red-carpet-heln-hunt-in-eco-friendly-hm/ (Accessed 28/08/13).

Korten, D. C. (1995) *When Corporations Rule the World*. Bloomfield, CT: Kumarian.

Krasner, S. (1983) *International Regimes*. Ithaca, NY: Cornell University Press.

Kroll, A. (2012) 'Are Walmart's Chinese Factories as Bad as Apple's?' *Mother Jones*. Available at: www.motherjones.com/environment/2012/03/walmart-china-sustainability-shadow-factories-greenwash?page=1 (Accessed 20/01/14).

Kuehls, T. (1996) *Beyond Sovereign Territory: The Space of Ecopolitics*. Minneapolis, MN: University of Minnesota Press.

Kuisma, M. (2008) 'Rights or Privilege? The Challenge of Globalization to the Values of Citizenship'. *Citizenship Studies* 12(6): 613–627.

Kukathas, C. (2007) *The Liberal Archipelago: A Theory of Diversity and Freedom*. Oxford: Oxford University Press.

Kurtz, H. (2007) 'Gender and Environmental Justice in Louisiana: Blurring the Boundaries of Public and Private Spheres'. *Gender, Place & Culture: A Journal of Feminist Geography* 14(4): 409–426.

Kusno, A. (2011) 'The Green Governmentality in an Indonesian Metropolis'. *Singapore Journal of Tropical Geography* 32(3): 314–331.

Kymlicka, W. (1995) *Multicultural Citizenship: A Liberal Theory of Minority Rights*. New York: Oxford University Press.

Kymlicka, W. (2010) 'Testing the Liberal Multiculturalist Hypothesis: Normative Theories and Social Science Evidence'. *Canadian Journal of Political Science* 43(2): 257–271.

Kymlicka, W. (2012) *Multiculturalism: Success, Failure, and the Future*. Washington, DC: Migration Policy Institute.

Kymlicka, W. and S. Donaldson (2014) 'Animals and the Frontiers of Citizenship'. *Oxford Journal of Legal Studies* 34(2): 201–219.

Kymlicka, W. and W. Norman (1994) 'Return of the Citizen: A Survey of Recent Work on Citizenship Theory'. *Ethics* 104(2): 352–381.

Læssøe, J. (2007) 'Participation and Sustainable Development: The Post-ecologist Transformation of Citizens Involvement in Denmark'. *Environmental Politics* 16(2): 231–250.

Lappé, A. M. and F. M. Lappé (2004) 'The Genius of Wangari Maathai'. *The New York Times*, 14 October 2004. Available at: ww.nytimes.com/2004/10/13/opinion/13iht-edlappe.html?_r=0 (Accessed 10/10/12).

Larner, W. and W. Walters (2004) 'Globalization as Governmentality'. *Alternatives* 29(5): 495–514.

Latour, B. (1993) *We Have Never Been Modern*. Translated by Catherine Porter. Cambridge, MA: Harvard University Press.

Latour, B. (2004) *The Politics of Nature: How to Bring the Sciences into Democracy*. Translated by Catherine Porter. Cambridge, MA: Harvard University Press.

Latta, A. (2007a) 'Citizenship and the Politics of Nature: The Case of Chile's Alto Bío Bío'. *Citizenship Studies* 11(3): 229–246.

Latta, A. (2007b) 'Locating Democratic Politics in Ecological Citizenship'. *Environmental Politics* 16(3): 377–393.

Latta, A. and N. Garside (2005) 'Perspectives on Ecological Citizenship: An Introduction'. *Environments Journal* 33(3): 1–8.

Latta, A. and H. Wittman (2010) 'Environment and Citizenship in Latin America: A New Paradigm for Theory and Practice'. *European Review of Latin American and Caribbean Studies* 89: 107–116.

Latta, A. and H. Wittman (eds.) (2012) *Environment and Citizenship in Latin America: Natures, Subjects and Struggles*. New York: Berghahn.

Law, W. W. (2011) *Citizenship and Citizenship Education in a Global Age: Politics, Policies and Practices in China*. New York: Peter Lang.

Lee, K., W. Oh and N. Kim (2013) 'Social Media for Socially Responsible Firms: Analysis of Fortune 500's Twitter Profiles and their CSR/CSIR Ratings'. *Journal of Business Ethics* 118(4): 791–806.

Leffers, D. and P. Ballamingie (2013) 'Governmentality, Environmental Subjectivity, and Urban Intensification'. *Local Environment* 18(2): 134–151.

Leopold, A. (1949) *A Sand County Almanac and Sketches Here and There.* Oxford: Oxford University Press.

Lester, L. (2010) *Media and the Environment: Conflict, Politics and the News.* Cambridge: Polity Press.

Lewis, T. (2012) '"There Grows the Neighborhood": Green Citizenship, Creativity and Life Politics on Eco-TV'. *International Journal of Cultural Studies* 15(3): 315–326.

Linklater, A. (1998) 'Cosmopolitan Citizenship'. *Citizenship Studies* 2(1): 23–41.

Linklater, A. (2007) *Critical Theory and World Politics: Citizenship, Sovereignty and Humanity.* London: Routledge.

Lister, R. (1997a) *Citizenship: Feminist Perspectives.* 2nd edition. New York: New York University Press.

Lister, R. (1997b) 'Citizenship: Towards a Feminist Synthesis'. *Feminist Review* 57(1): 28–48.

Lister, R. (1998) 'Citizenship and Difference: Towards a Differentiated Universalism'. *European Journal of Social Theory* 1(1): 71–90.

Lister, R. (2007) 'Inclusive Citizenship: Realizing the Potential'. *Citizenship Studies* 11(1): 49–61.

Lister, R. (2008) 'Inclusive Citizenship, Gender and Poverty: Some Implications for Education for Citizenship'. *Citizenship Teaching and Learning* 4(1): 3–19.

Locke, S. (2009) 'Environmental Education for Democracy and Social Justice in Costa Rica'. *International Research in Geographical and Environmental Education* 18(2): 97–110.

Lockie, S. (2009) 'Responsibility and Agency within Alternative Food Networks: Assembling the "Citizen Consumer"'. *Agriculture and Human Values* 26(3): 193–201.

Lockwood, M. (2010) 'A Tale of Two Milibands: From Environmental Citizenship to the Politics of the Common Good'. *The Political Quarterly* 81(4): 545–553.

Lomborg, B. (2001) *The Skeptical Environmentalist: Measuring the Real State of the World.* Cambridge: Cambridge University Press.

Longhurst, J. (2010) *Citizen Environmentalists.* Hanover, NE: University Press of New England.

Lorimer, J. (2010) 'International Conservation "Volunteering" and the Geographies of Global Environmental Citizenship'. *Political Geography* 29(6): 311–322.

Luke, T. W. (1997) *Ecocritique: Contesting the Politics of Nature, Economy, and Culture.* Minneapolis, MN: University of Minnesota Press.

Lyon, T. P. and A. W. Montgomery (2013) 'Tweetjacked: The Impact of Social

Media on Corporate Greenwash'. *Journal of Business Ethics* 118(4): 747–757.

Maathai, W. (2004a) *The Green Belt Movement: Sharing the Approach and the Experience.* New York: Lantern Books.

Maathai, W. (2004b) Nobel Lecture. Nobelprize.org, 10 December 2004. Available at: www.nobelprize.org/nobel_prizes/peace/laureates/2004/maathai-lecture-text.html (Accessed 12/07/12).

MacDonald, C. (2011) 'The Big Green Buyout'. *EMagazine*, 11 July 2011. Available at: www.emagazine.com/magazine/the-big-green-buyout (Accessed 20/02/13).

MacGregor, S. (2004a) 'From Care to Citizenship: Calling Ecofeminism Back to Politics'. *Ethics and the Environment* 9(1): 57–84.

MacGregor, S. (2004b) 'Reading the Earth Charter: Cosmopolitan Environmental Citizenship or Light Green Politics as Usual?' *Ethics, Place and Environment* 7(1–2): 85–96.

MacGregor, S. (2006a) *Beyond Mothering Earth: Ecological Citizenship and the Politics of Care.* Vancouver, BC: University of British Columbia Press.

MacGregor, S. (2006b) 'No Sustainability without Justice: A Feminist Critique of Environmental Citizenship', in A. Dobson and D. Bell (eds.), *Environmental Citizenship.* Cambridge, MA: The MIT Press, 101–126.

MacGregor, S. (2011) 'Citizenship and Care', in T. Fitzpatrick (ed.), *Understanding the Environment and Social Policy.* Bristol: The Policy Press, 271–289.

Macrory, R. (1996) 'Environmental Citizenship and the Law: Repairing the European Road'. *Journal of Environmental Law* 8(2): 220–235.

Maiangwa, B. and D. Agbiboa (2013) 'Oil Multinational Corporations, Environmental Responsibility and Turbulent Peace in the Niger Delta'. *Africa Spectrum* 48(2): 71–83.

Maignan, I. and O. C. Ferrell (2001) 'Corporate Citizenship as a Marketing Instrument: Concepts, Evidence and Research Directions'. *European Journal of Marketing* 35(3/4): 457–484.

Máiz, R. (2011) 'Igualdad, sustentabilidad y ciudadanía ecológica'. *Foro Interno* 11: 13–43.

Maniates, M. F. (2001) 'Individualization: Plant a Tree, Buy a Bike, Save the World?' *Global Environmental Politics* 1(3): 31–52.

Manville, P. H. (1990) *The Origins of Citizenship in Ancient Athens.* Princeton, NJ: Princeton University Press.

Mao, D. (2014) 'An Overview of the Green Movement in China', in T. Doyle and S. MacGregor (eds.), *Environmental Movements around the World: Shades of Green in Politics and Culture.* Santa Barbara, CA: Praeger, 207–234.

Marshall, P. (2008) *Demanding the Impossible: A History of Anarchism.* London: Harper Perennial.

Marshall, T. H. (1950) *Citizenship and Social Class and Other Essays.* Cambridge: Cambridge University Press.

Martínez-Alier, J. (2002) *The Environmentalism of the Poor: A Study of Ecological Conflicts and Valuation*. Northampton, MA: Edward Elgar Publishing.

Martinsson, J. and L. J. Lundqvist (2010) 'Ecological Citizenship: Coming Out "Clean" Without Turning "Green"'. *Environmental Politics* 19(4): 518–537.

Marzall, K. (2005). 'Environmental Extension: Promoting Ecological Citizenship'. *Environments Journal* 33(3): 65–77.

Maslow, A. H. (1943) 'A Theory of Human Motivation'. *Psychological Review* 50(4): 370–396.

Mason, A. (2009) 'Environmental Obligations and the Limits of Transnational Citizenship'. *Political Studies* 57(2): 280–297.

May, E. (2009) 'A Practical Environmental Education: Shrinking Ecological Footprints, Expanding Political Ones'. *Our Schools/Our Selves* 19(1): 17–24.

May, S. (2008) 'Ecological Citizenship and a Plan for Sustainable Development: Lessons from Huangbaiyu'. *City* 12(2): 237–244.

McChesney, R. (1999) *Rich Media, Poor Democracy: Communication Politics in Dubious Times*. Urbana, IL: University of Illinois Press.

McClintock, N. (2014) 'Radical, Reformist, and Garden-Variety Neoliberal: Coming to Terms with Urban Agriculture's Contradictions'. *Local Environment* 19(2): 147–171.

McDonough, W. and M. Braungart (2002) *Cradle to Cradle: Remaking the Way We Make Things*. New York: North Point Press.

Mead, L. (1986) *Beyond Entitlement: The Social Obligations of Citizenship*. New York: Free Press.

Meadows, D. H., D. L. Meadows, J. Randers and W. W. Behrens III (1972) *The Limits to Growth: A Report for the Club of Rome Project on the Predicament of Mankind*. London: Earth Island.

Mendes, C. (1989). *Fight for the Forest: Chico Mendes in His Own Words*. London: Latin American Bureau.

Merchant, C. (1996) *Earthcare: Women and the Environment*. New York: Routledge.

Merrett, C. D. (2007) 'Slow Food Movement', in P. Robbins (ed.), *Encyclopedia of Environment and Society*. Thousand Oaks, CA: SAGE Reference Online. Available at: http://dx.doi.org/10.4135/9781412953924

Merritt, A. and T. Stubbs (2012) 'Incentives to Promote Green Citizenship in UK Transition Towns'. *Development* 55(1): 96–103.

Miller, D. (1995) *On Nationality*. New York: Clarendon Press.

Miller, D. (2000) *Citizenship and National Identity*. Malden, MA: Polity Press.

Miller, T., N. Govil, J. McMurria, R. Maxwell and T. Wang (2005) *Global Hollywood 2*. London: BFI Publishing.

Mitchell, K. (2010) 'Top 10 Eco Education Trends: Blazing the Trail toward Sustainability'. *Alternatives Journal* 36(5): 6.

Monbiot, G. (2007) 'If Tesco and Wal-Mart Are Friends of the Earth, Are There Any Enemies Left?' *The Guardian*, 23 January 2007. Available at: www.monbiot.com/2007/01/23/the-new-friends-of-the-earth (Accessed 20/10/13).

Moore, N. (2011) 'Eco-Feminism and Rewriting the Ending of Feminism: From the Chipko Movement to Clayoquot Sound'. *Feminist Theory* 12(1): 3–21.

Morales, E. *et al.* (2011) *The Rights of Nature: The Case for a Universal Declaration of the Rights of Mother Earth*. San Francisco, CA: Council of Canadians, Fundación Pachamama.

Mrazek, R. (1996) 'Two Steps Forward, One Step Back: Developing an Environmentally Literate Citizenship in Canada'. *International Research in Geographical and Environmental Education* 5(2): 144–147.

Murphy, J. (2000) 'Ecological Modernisation'. *Geoforum* 31(1): 1–8.

Murray, R. L. and J. K. Heumann (2009) *Ecology and Popular Film: Cinema on the Edge*. Albany, NY: State University of New York Press.

Muthuki, J. (2006) 'Challenging Patriarchal Structures: Wangari Maathai and the Green Belt Movement in Kenya'. *Agenda* 69: 83–91.

Myers, G. (1995) '"The Power is Yours": Agency and Plot in "Captain Planet"', in C. Bazalgette and D. Buckingham (eds.), *In Front of the Children: Screen Entertainment and Young Audiences*. London: British Film Institute, 62–74.

Naess, A. (1973) 'The Shallow and the Deep. Long Range Ecology Movements: A Summary'. *Inquiry* 16(1): 95–100.

Naess, A. (1989) *Ecology, Community, and Lifestyle: Outline of an Ecosophy*. Cambridge: Cambridge University Press.

Nash, K. (2000) *Contemporary Political Sociology: Globalization, Politics, and Power*. Oxford: Blackwell Publishers.

Neuteleers, S. (2010) 'Institutions Versus Lifestyle: Do Citizens Have Environmental Duties in their Private Sphere?' *Environmental Politics* 19(4): 501–517.

Nicolosi, A. M. (2011) '"We Do Not Want Our Girls to Marry Foreigners": Gender, Race, and American Citizenship'. *NWSA Journal* 13(3): 1–21.

Nowak, A., H. Hale, J. Lindholm and E. Strausser (2009) 'The Story of Stuff: Increasing Environmental Citizenship'. *American Journal of Health Education* 40(6): 346–354.

Obach, B. (2009) 'Consumption, Ecological Footprints and Global Inequality: A Lesson in Individual and Structural Components of Environmental Problems'. *Teaching Sociology* 37(3): 294–300.

Okin, S. M. (1992) 'Is Multiculturalism Bad for Women?', in J. Cohen, M. Howard and M. C. Nussbaum (eds.), *Is Multiculturalism Bad for Women?* Princeton, NJ: Princeton University Press, 7–24.

Okin, S. M. (1998) [1991] 'Gender, the Public, and the Private', in A. Phillips (ed.) *Feminism and Politics*. Oxford: Oxford University Press, 116–142.

Oldfield, A. (1990) *Citizenship and Community*. London: Routledge.

Oleksy, E. H., J. Hearn and D. Golańska (eds.) (2011) *The Limits of Gendered Citizenship: Contexts and Complexities*. New York: Routledge.

Ophuls, W. (2011) *Plato's Revenge: Politics in the Age of Ecology*. Cambridge, MA: The MIT Press.

Opie, J. and N. Elliot (1996) 'Tracking the Elusive Jeremiad: The Rhetorical

Character of American Environmental Discourse', in J. G. Cantrill and
C. L. Oravec (eds.), *The Symbolic Earth: Discourse and Our Creation of the
Environment*. Lexington, KT: University of Kentucky, 9–37.

Opschoor, H. (2000) 'The Ecological Footprint: Measuring Rod or Metaphor?'
Ecological Economics 32(3): 363–365.

Orlove, B., R. Taddei, G. Podestá and K. Broad (2011) 'Environmental
Citizenship in Latin America: Climate, Intermediate Organizations and
Political Subjects'. *Latin American Research Review* 46: 115–149.

Ouellette, L. and J. Hay (2008) *Better Living through Reality TV: Television and
Post-welfare Citizenship*. Malden, MA: Blackwell Publishing.

Özen, S. and F. Küskü (2009) 'Corporate Environmental Citizenship Variation
in Developing Countries: An Institutional Framework'. *Journal of Business
Ethics* 89(2): 297–313.

Palacios, J. J. (2004) 'Corporate Citizenship and Social Responsibility in a
Globalized World'. *Citizenship Studies* 8(4): 383–402.

Pardy, B. (2005) 'Environmental Law and the Paradox of Ecological Citizenship:
The Case for Environmental Libertarianism'. *Environments Journal* 33(3):
25–36.

Parekh, B. (2000) *Rethinking Multiculturalism: Cultural Diversity and Political
Theory*. Basingstoke: Macmillan.

Parker, G. (1999) 'The Role of the Consumer-Citizen in Environmental Protest in
the 1990s'. *Space & Polity* 3(1): 67–83.

Partridge, D. J. (2011) 'Activist Capitalism and Supply-Chain Citizenship:
Producing Ethical Regimes and Ready-to-Wear Clothes'. *Current
Anthropology* 52 (Supp 3): S97–S111.

Pateman, C. (1989) *The Disorder of Women: Democracy, Feminism and
Political Theory*. Cambridge: Polity Press.

Paulson, N., A. Doolittle, A. Ladati, M. Welsh-Devine and P. Pena (2012)
'Indigenous Peoples' Participation in Global Conservation: Looking Beyond
Headdresses and Face Paint'. *Environmental Values* 21(3): 255–276.

Peeples, J. A. and K. M. DeLuca (2006) 'The Truth of the Matter: Motherhood,
Community and Environmental Justice'. *Women's Studies in Communication*
29(1): 59–87.

Pepper, D. (1996) *Modern Environmentalism: An Introduction*. London:
Routledge.

Peritore, N. P. (1999) *Third World Environmentalism: Case Studies from the
Global South*. Gainesville, FL: University Press of Florida.

Peter, J. and A. Wolper (eds.) (1995) *Women's Rights, Human Rights:
International Feminist Perspectives*. London: Routledge.

Pezzullo, P. C. (2011) 'Contextualizing Boycotts and Buycotts: The Impure
Politics of Consumer-Based Advocacy in an Age of Global Ecological Crises'.
Communication and Critical/Cultural Studies 8(2): 124–145.

Phillips, A. (1991) *Engendering Democracy*. Cambridge: Polity Press.

Phillips, A. (1993) *Democracy and Difference*. Cambridge: Polity Press.

Phillips, A. (2007) *Multiculturalism without Culture*. Princeton, NJ: Princeton University Press.

Phillips, C. (2005) 'Cultivating Practices: Saving Seed as Green Citizenship?' *Environments Journal* 33(3): 37–49.

Pierre, J. and B. G. Peters (2000) *Governance, Politics and the State*. Basingstoke: Macmillan.

Plumwood, V. (1992) 'Feminism and Ecofeminism: Beyond the Dualistic Assumptions of Women, Men, and Nature'. *The Ecologist* 22(1): 8–13.

Pocock, J. G. A. (1992) 'The Ideal of Citizenship Since Classical Times'. *Queen's Quarterly* 99(4): 35–55.

Pogge, T. W. (2002) *World Poverty and Human Rights: Cosmopolitan Responsibilities and Reforms*. Cambridge: Polity Press.

Ponthiere, G. (2009) 'The Ecological Footprint: An Exhibit at an Intergenerational Trial?' *Environment, Development and Sustainability* 11(4): 677–694.

Popović, N. (1996) 'In Pursuit of Environmental Human Rights: Commentary on the Draft Declaration of Principles on Human Rights and the Environment'. *Columbia Human Rights Law Review* 27(3): 487–603.

Postma, D. W. (2006) *Why Care for Nature?: In Search of an Ethical Framework for Environmental Responsibility and Education*. Dordrecht: Springer.

Prakash, A. (2000) *Greening the Firm: The Politics of Corporate Environmentalism*. New York: Cambridge University Press.

Presbey, G. M. (2013) 'Women's Empowerment: The Insights of Wangari Maathai'. *Journal of Global Ethics* 9(3): 277–292.

Putnam, R. (2000) *Bowling Alone: The Collapse and Revival of American Community*. New York: Simon & Schuster.

Pykett, J., P. Cloke, C. Barnett, N. Clarke and A. Malpass (2010) 'Learning to Be Global Citizens: The Rationalities of Fair-Trade Education'. *Environment and Planning D: Society and Space* 28(3): 487–508.

Rabinow, P. (ed.) (1991) *The Foucault Reader: An Introduction to Foucault's Thought*. London: Penguin.

Ramos, A. (2003) 'The Special (or Specious?) Status of Brazilian Indians'. *Citizenship Studies* 7(4): 401–420.

Rangan, H. (2000) *Of Myths and Movements: Rewriting Chipko into Himalayan History*. London: Verso.

Rasch, E. D. (2012) 'Transformations in Citizenship: Local Resistance against Mining Projects in Huehuetenango (Guatemala)'. *Journal of Developing Societies* 28(2): 159–184.

Raterman, T. (2012) 'Bearing the Weight of the World: On the Extent of an Individual's Environmental Responsibility'. *Environmental Values* 21(4): 417–436.

Redclift, M. and C. Sage (1998) 'Global Environmental Change and Global Inequality: North/South Perspective'. *International Sociology* 13(4): 499–516.

Rees, W. E. (2000) 'Eco-Footprint Analysis: Merits and Brickbats'. *Ecological Economics* 32(3): 371–374.

Reid, H. (2000) 'Embodying Ecological Citizenship: Rethinking the Politics of Grassroots Globalization in the United States'. *Alternatives: Global, Local, Political* 25(4): 439–467.

Reynolds, M., C. Blackmore and M. J. Smith (eds.) (2009) *The Environmental Responsibility Reader*. London: Zed Books.

Richardson, D. (1988) 'Sexuality and Citizenship'. *Sociology* 32(1): 83–100.

Riesenberg, P. (1992) *Citizenship in the Western Tradition: Plato to Rousseau*. Chapel Hill, NC: The University of North Carolina Press.

Riley, D. (1992) 'Citizenship and the Welfare State', in J. Allen, P. Braham and P. Lewis (eds.), *Political and Economic Forms of Modernity*. Cambridge: Polity Press.

Roach, B. (2007) 'Corporate Power in a Global Economy'. Medford, CT: Global Development and Environment Institute, Tufts University. Available at: www.ase.tufts.edu/gdae/education_materials/modules/Corporate_Power_in_a_Global_Economy.pdf (Accessed 20/04/12).

Roberts, J. (2011) *Environment and Policy*. 2nd edition. New York: Routledge.

Roche, M. (1992) *Rethinking Citizenship: Welfare, Ideology and Change in Modern Society*. Cambridge: Polity Press.

Rondinelli, D. A. and M. A. Berry (2000) 'Environmental Citizenship in Multinational Corporations: Social Responsibility and Sustainable Development'. *European Management Journal* 18(1): 70–84.

Rootes, C. (1999) 'Acting Globally, Thinking Locally? Prospects for a Global Environmental Movement'. *Environmental Politics* 8(1): 290–310.

Rose, N. (1996) 'Governing "Advanced" Liberal Democracies', in A. Barry, T. Osbourne and N. Rose (eds.), *Foucault and Political Reason: Liberalism, Neo-Liberalism and Rationalities of Government*. Chicago, IL: University of Chicago Press, 37–64.

Rose, N. (1999) *Powers of Freedom: Reframing Political Thought*. Cambridge: Cambridge University Press.

Ross, N. (2010) 'Is Walmart's Environmental Program a Substitute for Government Policy?' *Biodiversity: Alternative Journal* 36(6): 20–21.

Rowse, T. (2000) 'Indigenous Citizenship', in W. Hudson and J. Kane (eds.), *Rethinking Australian Citizenship*. Cambridge: Cambridge University Press, 86–98.

Ruddick, S. (1989) *Maternal Thinking: Towards a Politics of Peace*. London: Women's Press.

Ruggie, J. G. (1998) *Constructing the World Polity: Essays on International Institutionalisation*. London: Routledge.

Rumpala, Y. (2011) '"Sustainable Consumption" as a New Phase in a Governmentalization of Consumption'. *Theory and Society* 40(6): 669–699.

Sage, C. (2012) *Environment and Food*. London: Routledge.

Sagoff, M. (1988) *The Economy of the Earth: Philosophy, Law, and the Environment*. Cambridge: Cambridge University Press.

Sagoff, M. (2008) *The Economy of the Earth: Philosophy, Law, and the Environment*. 2nd edition. Cambridge: Cambridge University Press.

Sáiz, A. V. (2005) 'Globalization, Cosmopolitanism and Ecological Citizenship'. *Environmental Politics* 14(2): 163–178.

Sandel, M. (2012) *What Money Can't Buy: The Moral Limits of Markets*. London: Allen Lane.

Sandilands, C. (2000) 'Raising your Hand in the Council of all Beings: Ecofeminism and Citizenship'. *Ethics and the Environment* 4(2): 219–233.

Sarre, P. and S. Brown (1996) 'Changing Attitudes to Nature', in P. Sarre and A. Reddish (eds.), *Environment and Society*. 2nd edition. London: The Open University Press, 88–120.

Sauvé, L. (2005) 'Currents in Environmental Education: Mapping a Complex and Evolving Pedagogical Field'. *Canadian Journal of Environmental Education* 10: 11–37.

Scammell, M. (2000) 'The Internet and Civic Engagement: The Age of the Citizen-Consumer'. *Political Communication* 17(4): 351–355.

Schell, O. (2011) 'How Walmart Is Changing China'. *The Atlantic Magazine*, 26 October 2011. Available at: www.theatlantic.com/magazine/archive/2011/12/how-walmart-is-changing-china/308709/ (Accessed 20/01/14).

Schmink, M. (2011) 'Forest Citizens: Changing Life Conditions and Social Identities in the Land of the Rubber Tappers'. *Latin American Research Review* 46(4): 141–158.

Schmitt, M. (2005) 'We're All Environmentalists Now'. *The American Prospect*, 18 September 2005. Available at: http://prospect.or/article/were-all-enviromentalists-now (Accessed 06/12/12).

Schudson, M. (2007) 'Citizens, Consumers, and the Good Society'. *The Annals of the American Academy of Political and Social Science* 611(1): 236–249.

Sejersen, T. B. (2008) '"I Vow to Thee My Countries" – The Expansion of Dual Citizenship in the 21st Century'. *International Migration Review* 42(3): 523–549.

Sending, O. J. and I. B. Neumann (2006) 'Governance to Governmentality: Analyzing NGOs, States, and Power'. *International Studies Quarterly* 50(3): 651–672.

Serrès, M. (1995) [1990] *The Natural Contract*. Translated by E. MacArthur and W. Paulson. Ann Arbor, MI: University of Michigan Press.

Seyfang, G. (2005) 'Shopping for Sustainability: Can Sustainable Consumption Promote Ecological Citizenship?' *Environmental Politics* 14(2): 290–306.

Seyfang, G. (2006) 'Ecological Citizenship and Sustainable Consumption: Examining Local Organic Food Networks'. *Journal of Rural Studies* 22(4): 383–395.

Shachar, A. (2009) *The Birthright Lottery: Citizenship and Global Inequality.* Cambridge, MA: Harvard University Press.

Shafir, G. (1998) *The Citizenship Debates: A Reader.* Minneapolis, MN: University of Minnesota Press.

Shafir, G. and A. Brysk (2006) 'The Globalization of Rights: From Citizenship to Human Rights'. *Citizenship Studies* 10(3): 275–287.

Shaver, S. (2004) 'Welfare, Equality and Globalisation', in K. Horton and H. Patapan (eds.), *Globalisation and Equality.* London: Routledge, 95–113.

Shaw, D., T. Newholm and R. Dickinson (2006) 'Consumption as Voting: An Exploration of Consumer Empowerment'. *European Journal of Marketing* 40(9/10): 1049–1067.

Shellenberger, M. and T. Nordhaus (2004) 'The Death of Environmentalism: Global Warming Politics in a Post-Environmental World'. Available at: www.thebreakthrough.org/images/Death_of_Environmentalism.pdf (Accessed 20/01/14).

Shiva, V. (1997) *Biopiracy: The Plunder of Nature and Knowledge.* Cambridge, MA: South End Press.

Shiva, V. (2001) *Protect or Plunder? Understanding Intellectual Property Rights.* London: Zed Books.

Shiva, V. (2005) *Earth Democracy: Justice, Sustainability, and Peace.* London: Zed Books.

Sigler, C. (1994) 'Wonderland to Wasteland: Toward Historicizing Environmental Activism in Children's Literature'. *Children's Literature Association Quarterly* 19(4): 148–153.

Siim, B. (2000) *Gender and Citizenship: Politics and Agency in France, Britain and Denmark.* Cambridge: Cambridge University Press.

Simon, J. L. and H. Khan (eds.) (1984) *The Resourceful Earth: A Response to 'Global 2000'.* New York: Basil Blackwell.

Smith, G. and D. Williams (2000) 'Ecological Education: Extending the Definition of Environmental Education'. *Australian Journal of Environmental Education* 15–16: 139–146.

Smith, M. (1998) *The Myth of Green Marketing: Tending Our Goats at the Edge of Apocalypse.* Toronto: University of Toronto Press.

Smith, M. (2005) 'Obligation and Ecological Citizenship'. *Environments Journal* 33(3): 9–23.

Smith, M. J. (1998) *Ecologism: Towards Ecological Citizenship.* Buckingham: Open University Press.

Smith, M. J. and P. Pangsapa (2008) *Environment and Citizenship: Integrating Justice, Responsibility and Civic Engagement.* London: Zed Books.

Smith, M. J. and E. Parsons (2012) 'Animating Child Activism: Environmentalism and Class Politics in Ghibli's *Princess Mononoke* (1997) and Fox's *Fern Gully* (1992)'. *Continuum: Journal of Media and Cultural Studies* 26(1): 25–37.

Smith, R. M. (2002) 'Modern Citizenship', in E. F. Isin and B. S. Turner (eds.), *Handbook of Citizenship Studies*. London: Sage, 105–115.

Soper, K. (1995) *What Is Nature?* Oxford: Blackwell.

Soper, K. (2004) 'Rethinking the Good Life: Consumer as Citizen'. *Capitalism, Nature, Socialism* 15(3): 111–116.

Soper, K. (2007) 'Rethinking the "Good Life": The Citizenship Dimension of Consumer Disaffection with Consumerism'. *Journal of Consumer Culture* 7(2): 205–229.

Soper, K. and F. Trentmann (eds.) (2008) *Citizenship and Consumption*. Basingstoke: Palgrave Macmillan.

Sørensen, E. (2002) 'Democratic Theory and Network Governance'. *Administrative Theory and Praxis* 24(4): 693–720.

Spash, C. L. (2012a) 'Green Economy, Red Herring'. *Environmental Values* 21(2): 95–99.

Spash, C. L. (2012b) 'Response and Responsibility'. *Environmental Values* 21(4): 391–396.

Spence, M. (1996) 'Dispossessing the Wilderness: Yosemite Indians and the National Park Ideal, 1864–1930'. *Pacific Historical Review* 65(1): 27–59.

Stapp, W. B. *et al.* (1969) 'The Concept of Environmental Education'. *The Journal of Environmental Education* 1(1): 30–31.

Starosielski, N. (2011) '"Movements that Are Drawn": A History of Environmental Animation from *The Lorax* to *FernGully* to *Avatar*'. *The International Communication Gazette* 73(1–2): 145–163.

Stavenhagen, R. (2013) 'How to Decolonize Indigenous Rights'. *Latin American and Caribbean Ethnic Studies* 8(1): 97–102.

Stein, R. (ed.) (2004) *New Perspectives on Environmental Justice: Gender, Sexuality and Activism*. New Brunswick, NJ: Rutgers University Press.

Stephens, B. (2002) 'The Amorality of Profit: Transnational Corporations and Human Rights'. *Berkeley Journal of International Law* 20(1): 45–90.

Stevenson, N. (2002) 'Consumer Culture, Ecology and the Possibility of Cosmopolitan Citizenship'. *Consumption, Markets and Culture* 5(4): 305–319.

Stolle, D., M. Hooghe and M. Micheletti (2005) 'Politics in the Supermarket: Political Consumerism as a Form of Political Participation'. *International Political Science Review* 26(3): 245–269.

Stone, C. (1972) 'Should Trees Have Standing? Toward Legal Rights for Natural Objects'. *Southern California Law Review* 45: 450–501.

Sturgeon, N. (2009) *Environmentalism in Popular Culture: Gender, Race, Sexuality, and the Politics of the Natural*. Tucson, AZ: The University of Arizona Press.

Swallow, L. and J. Furniss (2011) 'Green Business: Reducing Carbon Footprint Cuts Costs and Provides Opportunities'. *Montana Business Quarterly* 49(2): 2–9.

Swanson, T. (ed.) (1995) *Intellectual Property Rights and Biodiversity Conservation: An Interdisciplinary Analysis of the Values of Medicinal Plants*. Cambridge: Cambridge University Press.

Szerszynski, B. (2007) 'The Post-ecologist Condition: Irony as Symptom and Cure'. *Environmental Politics* 16(2): 337–355.

Tarbotton, R. (2010) 'Has Earth Day Become Corporate Greenwash Day?' *The Huffington Post*, 20 April 2010. Available at: www.huffingtonpost.com/rebecca-tarbotton/has-earth-day-become-corp_b_548066.html (Accessed 24/03/13).

Tarrant, M. and K. Lyons (2012) 'The Effect of Short-Term Educational Travel Programs on Environmental Citizenship'. *Environmental Education Research* 18(3): 403–416.

Taylor, B. (2010) *Dark Green Religion: Nature Spirituality and the Planetary Future*. Berkeley, CA: University of California Press.

The Economist (2005a) The Union of Concerned Executives. [Special report: Corporate Social Responsibility]. 20 January 2005. Available at: www.economist.com/node/3555194/print (Accessed 30/09/13).

The Economist (2005b) The World According to CSR. [Special report: Corporate Social Responsibility]. 20 January 2005. Available at: www.economist.com/node/3555272/print (Accessed 30/09/13).

Thussu, D. K. (ed.) (1998) *Electronic Empires: Global Media and Local Resistance*. London: Arnold.

Thussu, D. K. (2007) *News as Entertainment: The Rise of Global Infotainment*. London: Sage.

Tilden, T. L. (2008) 'How Green Celebrities Helped Save Our Planet. A Plethora of Celebrities Took Pauses for Causes'. *The Daily Green*, 13 September 2008. Available at: www.thedailygreen.com/living-green/blogs/celebrities/green-celebrities-summer (Accessed 14/10/12).

Todd, A. M. (2002) 'Prime-Time Subversion: The Environmental Rhetoric of *The Simpsons*', in M. Meister and P. M. Japp (eds.), *Enviropop: Studies in Environmental Rhetoric and Popular Culture*. Westport, CT: Praeger, 63–80.

Torgerson, D. (2000) 'Farewell to the Green Movement? Political Action and the Green Public Sphere'. *Environmental Politics* 9(4): 1–19.

Trachtenberg, Z. (2010) 'Complex Green Citizenship and the Necessity of Judgement'. *Environmental Politics* 19(3): 339–355.

Travaline, K. and C. Hunold (2010) 'Urban Agriculture and Ecological Citizenship in Philadelphia'. *Local Environment* 15(6): 581–590.

Trentmann, F. (2007) 'Citizenship and Consumption'. *Journal of Consumer Culture* 7(2): 147–158.

Turner, B. (2011) 'Embodied Connections: Sustainability, Food Systems and Community Gardens'. *Local Environment* 16(6): 509–522.

Turner, B. S. (1986) *Citizenship and Capitalism: The Debate over Reformism*. London: Allen & Unwin.

UNEP (1972) Declaration of the United Nations Conference on the Human

Environment. Available at: www.unep.org/Documents.Multilingual/Default. asp?documentid=97&articleid=1503 (Accessed 16/01/13).

UNEP (1992a) Rio Declaration on Environment and Development. Available at: www.unep.org/Documents.Multilingual/Default.asp?documentid= 78&articleid=1163 (Accessed 16/01/13).

UNEP (1992b) Agenda 21. Available at: http://sustainabledevelopment.un.org/ content/documents/Agenda21.pdf (Accessed 16/01/13).

UNEP (1997) Global Environmental Outlook. Available at: www.unep.org/geo/ geo1/ch/ch3_9.htm (Accessed 30/01/14).

UNEP (2001) Project Document: Global Environmental Citizenship (GEC). Latin America and the Caribbean. Available at: www.thegef.org/gef/sites/ thegef.org/files/gef_prj_docs/GEFProjectDocuments/Biodiversity/Global% 20-%20Global%20Environment%20Citizenship/Project%20Document.pdf (Accessed 30/01/13).

UNEP (2002) Annual Report (Environmental Citizenship Program). Available at: www.unep.or.jp/ietc/Activity_Report/2002/13.asp (Accessed 16/01/13).

UNEP (2003) Shopping for a Better World. 2 June 2003. Available at: www.unep.org/Documents.Multilingual/Default.asp?DocumentID=321& ArticleID=4019 (Accessed 30/01/13).

UNEP (2007) Public Environmental Awareness and Education. Available at: www.unep.org/training/programmes/Instructor%20Version/Part_2/Activities/ Interest_Groups/Public_Awareness/Core/Public_Environmental_Awareness_ and_Education.pdf (Accessed 30/01/13).

UNEP (2009) Terminal Evaluation of the UNEP GF/5024-02-01 (4485) 'Global Environmental Citizenship (GEC)'. [Anne Fouillard – Evaluation Office] Available at: www.unep.org/eou/Portals/52/Reports/GEC%20Final%20 Report.pdf (Accessed 30/01/13).

UNEP (2011) Towards a Green Economy: Pathways to Sustainable Development and Poverty Eradication. Available at: www.unep.org/ greeneconomy/Portals/88/documents/ger/GER_synthesis_en.pdf (Accessed 22/02/2013).

UNEP (2012a) The Future We Want. Available at: www.uncsd2012.org/content/ documents/774futurewewant_english.pdf (Accessed 30/01/13).

UNEP (2012b) On the Road to Rio+20, Countries Accelerate Plans Green Economy Transition. Available at: www.unep.org/ecosystemmanagement/ News/PressRelease/tabid/426/language/en-US/Default.aspx?DocumentID=26 59&ArticleID=8943&Lang=en (Accessed 22/02/13).

UNEP (2014) Green Passport. Available at: www.unep.org/resourceefficiency/ Business/SectoralActivities/Tourism/Activities/GreenPassport/tabid/78823/ Default.aspx (Accessed 02/08/14).

UNESCO (1977) Tbilisi Declaration. Available at: www.gdrc.org/uem/ee/tbilisi. html (Accessed 10/02/13).

Upham, P. (2012) 'Environmental Citizens: Climate Pledger Attitudes and Micro-Generation Installation'. *Local Environment* 17(1): 75–91.

Valdivieso, J. (2005) 'Social Citizenship and the Environment'. *Environmental Politics* 14(2): 239–254.

Vale, R. and B. Vale (eds.) (2013) *Living Within a Fair Share Ecological Footprint*. London: Routledge.

Van den Bergh, J. C. J. M. and H. Verbruggen (1999) 'Spatial Sustainability, Trade and Indicators: An Evaluation of the "Ecological Footprint"'. *Ecological Economics* 29(1): 61–72.

Van Kooten, G. C. and E. H. Bulte (2000) 'The Ecological Footprint: Useful Science or Politics?' *Ecological Economics* 32(3): 385–389.

Van Steenbergen, B. (ed.) (1994) *The Condition of Citizenship*. London: Sage.

Vivanco, L. A. (2002) 'Seeing Green: Knowing and Saving the Environment on Film'. *Visual Anthropology* 104(4): 1195–1204.

Vivanco, L. A. (2004) 'The Work of Environmentalism in an Age of Televisual Adventures'. *Cultural Dynamics* 16(1): 5–27.

Vogel, D. (2005) *The Market for Virtue: The Potential and Limits of Corporate Social Responsibility*. Washington, DC: Brooking Institution Press.

Wackernagel, M. and W. Rees (1996) *Our Ecological Footprint: Reducing Human Impact on the Earth*. Gabriola Island, BC: New Society Publishers.

Wagner, A. (2004) 'Redefining Citizenship for the 21st Century: From the National Welfare State to the UN Global Compact'. *International Journal of Social Welfare* 13: 278–286.

Wake, S. J. and C. Eames (2013) 'Developing an "Ecology of Learning" within a School Sustainability Co-Design Project with Children in New Zealand'. *Local Environment* 18(3): 305–322.

Walby, S. (1994) 'Is Citizenship Gendered?' *Sociology* 28(2): 379–395.

Wallerstein, I. (2003) 'Citizens All? Citizens Some! The Making of the Citizen'. *Comparative Studies in Society and History* 45(4): 650–679.

Wapner, P. (1996) 'Toward a Meaningful Ecological Politics'. *Tikkun* 11(3): 21–26.

Warren, M. (2008) 'Drama, Not Doomsday'. *The Australian*, 28 August 2008. Available at: www.theaustralian.com.au/news/features/drama-not-doomsday/story-e6frg6z6-1111117318682 (Accessed 28/06/11).

Watts, M. (2002) 'Green Capitalism, Green Governmentality'. *American Behavioral Scientist* 45(9): 1313–1317.

WCED (1987) *Our Common Future*. Oxford: Oxford University Press.

Weis, T. (2013) 'The Meat of the Global Food Crisis'. *The Journal of Peasant Studies* 40(1): 65–85.

Westheimer, J. and J. Kahne (2004) 'What Kind of Citizen? The Politics of Educating for Democracy'. *American Educational Research Journal* 41(2): 237–269.

Whelan, G., J. Moon and B. Grant (2013) 'Corporations and Citizenship Arenas in the Age of Social Media'. *Journal of Business Ethics* 118(4): 777–790.

White, T. (2007) 'Sharing Resources: The Global Distribution of the Ecological Footprint'. *Ecological Economics* 64(2): 402–410.

Whitehead, M. (2011) 'Environment Inc. and Panda Logos'. *Environmental Values* 20(1): 1–5.

Whitley, D. (2008) *The Idea of Nature in Disney Animation*. Farnham, Surrey: Ashgate.

Wilkinson, D. (2002) *Environment and Law*. London: Routledge.

Willard, B. (2002) *The Sustainability Advantage: Seven Business Case Benefits of a Triple Bottom Line*. Gabriola Island, BC: New Society Publishers.

William, W. (2002) 'Citizenship Questions and Environmental Crisis in the Niger Delta: A Critical Reflection'. *Nordic Journal of African Studies* 11(3): 377–392.

Williams, G. (2008) 'Cosmopolitanism and the French Anti-GM Movement'. *Nature and Culture* 3(1): 115–133.

Winter, G. (2012) 'Environmental Governance in Germany', in M. Alberton and F. Palermo (eds.), *Environmental Protection in Multi-Layered Systems: Comparative Lessons from the Water Sector*. Leiden: Martinus Nijhoff Publishers, 55–81.

Wissenburg, M. (1998) *Green Liberalism: The Free and the Green Society*. London: UCL Press.

Wittman, H. (2009a) 'Reworking the Metabolic Rift: La Vía Campesina, Agrarian Citizenship, and Food Sovereignty'. *The Journal of Peasant Studies* 36(4): 805–826.

Wittman, H. (2009b) 'Reframing Agrarian Citizenship: Land, Life and Power in Brazil'. *Journal of Rural Studies* 25(1): 120–130.

Wittman, H. (2009c) 'Agrarian Reform and the Environment: Fostering Ecological Citizenship in Mato Grosso, Brazil'. *Canadian Journal of Development Studies* 29(3–4): 281–298.

WSSD (2002) Johannesburg Declaration on Sustainable Development. Available at: www.un-documents.net/jburgdec.htm (Accessed 30/01/13).

WWF (2012) Living Planet Report 2012. Available at: http://wwf.panda.org/about_our_earth/all_publications/living_planet_report/2012_lpr/ (Accessed 03/12/14).

WWF-Malaysia and Partners (2008) Environmental Citizenship: A Report on Emerging Perspectives in Malaysia. Available at: https://d1kjvfsq8j7onh.cloudfront.net/downloads/environmental_citizenship_study_report_170510.pdf (Accessed 20/01/13).

Yanitsky, O. N. (2012) 'From Nature to Politics: The Russian Environmental Movement 1960–2010'. *Environmental Politics* 21(6): 922–940.

Young, I. M. (1989) 'Polity and Group Difference: A Critique of the Ideal of Universal Citizenship'. *Ethics* 99(2): 250–274.

Young, I. M. (1990) *Justice and the Politics of Difference*. Princeton, NJ: Princeton University Press.

Yu, J., K. R. Coulson, J. X. Zhou, H. J. Wen and Q. Zhao (2011) 'Communicating Corporate Environmental Citizenship: An Examination of Fortune 500 Web Sites'. *Journal of Internet Commerce* 10(3): 193–207.

Yuval-Davis, N. (1997) *Gender and Nation*. London: Sage.

Yuval-Davis, N. (2011) *The Politics of Belonging: Intersectional Contestations*. London: Sage.

Zadek, S. (2004) 'The Path to Corporate Responsibility'. *Harvard Business Review* 82(12): 125–132.

Index